ENVIRONMENTAL AND TECHNOLOGY POLICY IN EUROPE
Technological Innovation and Policy Integration

ENVIRONMENT & POLICY

VOLUME 38

The titles published in this series are listed at the end of this volume.

Environmental and Technology Policy in Europe

Technological Innovation and Policy Integration

Edited by

Geerten J.I. Schrama

Center for Clean Technology and Environmental Policy,
University of Twente, The Netherlands

and

Sabine Sedlacek

Department of Environmental Economics and Management,
Vienna University of Economics and Business Administration, Austria

The ENVINNO research project, "Towards an Integration of Environmental and Ecology-Oriented Technology Policy: Stimulus and Response in Environment Related Innovation Networks", has been funded by:

   EUROPEAN COMMISSION
DIRECTORATE-GENERAL FOR RESEARCH

Targeted Social and Economic Research (TSER) Programme

Under contract-number SOE1-CT98-1105

KLUWER ACADEMIC PUBLISHERS
DORDRECHT / BOSTON / LONDON

A C.I.P. Catalogue record for this book is available from the Library of Congress.

ISBN 1-4020-1583-6

Published by Kluwer Academic Publishers,
P.O. Box 17, 3300 AA Dordrecht, The Netherlands.

Sold and distributed in North, Central and South America
by Kluwer Academic Publishers,
101 Philip Drive, Norwell, MA 02061, U.S.A.

In all other countries, sold and distributed
by Kluwer Academic Publishers,
P.O. Box 322, 3300 AH Dordrecht, The Netherlands.

Printed on acid-free paper

Printed in the Netherlands.

Table of contents

Chapter 4 Environmental Policy and Environment-oriented Technology Policy in Germany 97

Jobst Conrad

Chapter 5 Environmental Policy and Environment-oriented Technology Policy in the Netherlands 125

Peter S. Hofman and Geerten J.I. Schrama

Chapter 6 Environmental Policy and Environment-oriented Technology Policy in Spain 163

Jose Carlos Cuerda, Maria Jose Fernandez, Juan Larrañeta,
Susana Muñoz, Florencio Sanchez, and Carmen Velez

Chapter 7 Environmental Policy and Environment-oriented Technology Policy in the United Kingdom 197

John F. Grant and Nigel D. Mortimer

Chapter 8 Synthesis   **225**
Sabine Sedlacek and Geerten J.I. Schrama

Preface

This book contains six studies on various national environmental policies and environment-oriented technology policy systems in Austria, Denmark, Germany, the Netherlands, Spain, and the United Kingdom, sandwiched between this introductory and a concluding chapter. These studies were conducted as part of the ENVINNO research project, *"Towards an Integration of Environmental and Ecology-Oriented Technology Policy: Stimulus and Response in Environment Related Innovation Networks"*, which formed part of the Targeted Social and Economic Research (TSER) Programme of Directorate-General XII of the European Commission, now Directorate-General for Research.[1]

We like to thank Mrs. Genevieve Zdrojewski of GD Directorate-General Research for her kind support of our research project.

The project was carried out between 1998 and 2001 by research teams from the six countries. The co-ordinating institute was the Department of Environmental Economics and Management at the Vienna University of Economics and Business Administration.[2]

At this place we want to mention all researchers involved in the ENVINNO project and we want to thank them all for their contributions to this book and the project and for the good time we have had performing the project and meeting each other at regular intervals in Vienna (A), Enschede (NL), Berlin (D), and Sevilla (E).

Department of Environmental Economics and Management at the Vienna University of Economics and Business Administration in Austria:
- Univ. Prof. Dr. Uwe Schubert,
- Mag. Judith Köck,
- Mag. Jürgen Mellitzer,

---

[1] Under contract-number SOE1-CT98-1105.
[2] The project website is http://www.wu-wien.ac.at/wwwu/institute/iuw/ENVINNO.

- Dr. Sabine Sedlacek,
- Dr. Peter Townroe,
- Dr. Andreas Zerlauth.

Environment Policy Research Unit at the Free University of Berlin in Germany:
- Dr. Jobst Conrad (currently at the Center for Environmental Research Leipzig-Halle),
- Prof. Dr. Martin Jänicke.

Center for Clean Technology and Environmental Policy at the University of Twente in the Netherlands:
- Drs. Peter S. Hofman, M.A., MBA,
- Dr. Geerten J.I. Schrama.

Department of Environment, Technology and Social Studies at the University of Roskilde in Denmark:
- Dr. Ole Erik Hansen,
- Dr. Jesper Holm,
- Dr. Bent Søndergaard.

Institute for Regional Development, Fundación Universitaria in Sevilla, Spain:
- Prof. Dr. Juan Larrañeta Astola (currently at the Escuela Superior de Ingenieros de Sevilla),
- Dr. Jose Carlos Cuerda Garcia-Junceda,
- Maria Jose Fernandez Lopez,
- Susana Muñoz Rouco,
- Florencio Sanchez Escobar,
- Carmen Velez Mendez.

Resources Research Unit at the Sheffield Hallam University, in the United Kingdom:
- Prof. Dr. Nigel D. Mortimer,
- John F. Grant.

Chapter 1

Introduction

GEERTEN J.I. SCHRAMA
Center for Clean Technology and Environmental Policy, University of Twente, the Netherlands

SABINE SEDLACEK
Department of Environmental Economics and Management, Vienna University of Economics and Business Administation, Austria

1.1   Background of the book

1.1.1   Subject and focus

It is clear that improvements to environmental conditions are dependent to a large extent on innovation by industry. Without going into detail on the debates about the relationship between innovation and environment, or whether technological innovations always lead to environmental improvements, one can stipulate a class of innovations that have positive effects on the environment that can be called, in line with generally established views, 'sustainable' or 'environment-oriented innovations'. This book is about public policy that is aimed at stimulating the adoption of environment-oriented innovations by industry, and it is based on a comparative analysis of the developments over recent decades in six EU member states.

The decision to innovate depends on many factors. At the level of the individual company, or the individual innovation project, one can distinguish a multitude of economic, social, and political factors that interact with each other. In terms of political factors, two policy fields are seen as the most relevant: environmental policy and technology policy (see figure 1.1:

1

Geerten J.I. Schrama and Sabine Sedlacek (eds.) Environmental and Technology Policy in Europe. Technological innovation and policy integration, 1-24. ©2003 Kluwer Academic Publishers. Printed in the Netherlands.

ENVINNO conceptual model). Instead of the individual company or innovation project, this book adopts the perspective of 'policy approaches' which are characteristics of particular countries in particular periods of time. While environment-oriented innovations are a standard element of the approaches used in the field of environmental policy, the environment and sustainability are not obvious goals of technology policy. Environmental policy, in a very general sense, concerns the state of the environment, which as such has a value. Stimulating sustainable innovation is instrumental in, and additional to, policy measures that aim to influence the behaviour of industry and other target groups as it effects the environment. Technology policy is aimed at stimulating research and development for new technologies, and the diffusion of these innovations, which does not have a value as such. The rationale behind technology policy is, in a very general sense, to support the competitiveness of the national economy. Often measures and programmes concerning technology policy have specific targets, such as specific regions, types of technology, or contributions to other policy fields. Sustainable innovations are frequently a specific target of technology policy measures and programmes. For those parts of technology policy to which this applies, we use the term 'environment-oriented technology policy' (ETP).

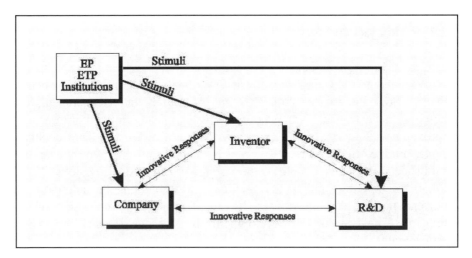

Figure 1.1 ENVINNO conceptual model

Environment-oriented technology policy addresses the problem of the reluctance of the private sector to engage in activities that would improve the efficiency of natural resource use, particularly with respect to environmental pollution. Strategies adopted to reduce the above-mentioned risks borne by the individual actors in innovation processes vary considerably, but there is a consensus that societal action is necessary to overcome the gridlock.

Environment policy and environment-oriented technology policy, like any other kind of public policy, is considered here as means for public authorities to affect the conduct of social actors or 'target groups'. The basic mechanism is that policy is executed through the application of policy instruments containing certain incentives for the target group. Policy instruments are effective to the extent that they succeed in getting the members of the target group to reconsider their conduct in accordance with the policy makers' intentions. Basically three models of social steering can be distinguished according to the type of policy incentives: economic, juridical, and social (c.f. Schrama, 1998). In the case of stimulating environment-oriented innovations through public policy, the companies that should apply the innovations are not always the first addressees of the policy incentives. In stead of direct steering, ways of indirect steering using intermediaries, such as the inventors and R&D organisations in figure 1.1, are involved.

Indirect steering is characteristic for the 'Stimulus-Organism-Response model', which is central to the ENVINNO research design. The model is developed by sociologists and adopted by business economists (e.g. Meffert, 1992; Schmalen and Pechtl, 1992; Brunner and Mellitzer, 1996) to study the effects of external and internal influences on the actions of companies. The core research issue of the ENVINNO project were the responses of companies and R&D organisations to stimuli provided by EP and ETP. In the context of innovation stimuli are seen as all those variables leading to innovative responses. Innovative responses are to be distinguished from routine responses and non-responses, and can be classified into three categories: technological, organisational, and marketing responses (e.g. Nelson, 1993; Schmalen and Pechtl, 1992).

By addressing the policy approaches in two different fields, and the way they stimulate the adoption of environment-oriented innovations by industry, one also touches on the issue of 'policy integration'. In Europe today, several approaches co-exist (Andersen and Liefferink, 1997) in which these two policy areas are largely integrated into a coherent policy system. In this book, policy integration is studied at the national level in terms of co-ordination of policy goals, the application of means, and advanced forms of inter-policy co-operation. These issues are addressed through an analysis of the way in which public policy has dealt with sustainable innovations by industry in recent decades and, in particular, the conditions that lead to the emergence of co-ordination between the environment and technology policy fields. In addition to the analysis at the national level, the issue of 'environmental integration' within the European Union is also addressed within this introductory chapter.

1.1.2   The ENVINNO research project

This book contains six studies on national environmental policies and environment-oriented technology policy systems in Austria, Denmark, Germany, the Netherlands, Spain, and the United Kingdom, sandwiched between this introductory and a concluding chapter. These studies were originally conducted as part of the ENVINNO research project, 'Towards an Integration of Environmental and Ecology-Oriented Technology Policy: Stimulus and Response in Environment Related Innovation Networks', which was carried out, between 1998 and 2001, by research teams from the six countries.

The principal aims of the ENVINNO project were to determine the effectiveness of the various environmental policy (EP) and environment-oriented technology policy (ETP) approaches to implement sustainable development goals, to identify areas of success and failure, and to formulate practical means for developing successful procedures in the future. The aims of the project can be summarised as follows:

– to conduct a policy analysis to establish the key features of the environmental policies and environment-oriented technology policies which are functioning in different EU Member States, to determine their effectiveness, and to select examples of effective policy instruments,
– to investigate, by means of case studies, the experiences of innovative companies that have been involved in the development and commercial exploitation of environmental technology, to draw conclusions about the success and failure of the current innovation processes, and to recommend measures for improving this process, and
– to contribute, where possible, to the development, implementation, and maintenance of practical methods for assisting the process of innovation in environmental technology in individual member states and across the European Union.

The country studies reported here are the results of the activities undertaken to meet the aims outlined under the first bullet point above.

1.1.3   Central concepts

In this section, the central concepts used in the book are clarified.

1.1.3.1   Innovation

Innovation, and in particular sustainable or environment-oriented innovation, is a core concept in this book. Innovation is a common word and its meaning is more or less evident. However, for a study such as this, one has to be specific about the meaning of this concept.

Firstly, we are considering innovation in industry, in the sense of the renewal of production processes or products. According to the usual classification used in innovation theory, one can distinguish between technological, organisational, and social innovations (e.g. Freeman, 1982; Kline and Rosenberg, 1986; Dosi et al., 1988; Feldman, 2000; Cooke et al., 2000). In line with the focus on product and process innovation, the research is focussed on technological innovations, with the term 'technological' referring to knowledge and expertise which is, at least partially, embodied in artefacts (e.g. Schrama, 1991).

Secondly, innovation here refers to particular processes that can be interpreted at various levels, such as the global level, the level of a particular country or economy, a particular industry, or a single company. The innovation process from the perspective of a single company can be understood as the search for a solution to a specific problem. The solution can be developed within the company or imported. The latter is referred to as 'diffusion': it is an innovation at the level of the company, but not at higher levels. Public policy that aims to stimulate environment-oriented innovations is directed at various levels: from the stimulation of companies to adopt new technologies, to the support for the R&D infrastructure at the national level. The ENVINNO project takes a network view of innovation processes, acknowledging the roles of all the actors, from R&D institutions to the individual company as 'end-user'. Public policy incentives can have an impact anywhere in the innovation network. However, our specific research interest is in the way these policy incentives eventually affect the adopting company, either directly or indirectly through a stimulus-response nexus.

Finally, innovation is usually characterised in terms of 'radical' versus 'incremental'. Radical innovations involve *"discontinuous change and the introduction of new technologies"*, whereas incremental innovations are often defined as *"gradual improvements of existing technologies"* (Murphy and Gouldson, 2000: 35). Recently this distinction has been extended with concepts such as 'transition' or 'breakthrough'. While the term radical refers to the extent to which the new technology replaces existing technology, or makes it obsolete; the concepts of 'transition' and 'breakthrough' refer to the wider context in which the technology is used. Acceptance and application of a new technology on a large scale may require substantial changes in context or system. The ENVINNO project has no restrictions in this respect, although both extremes of the continuum (simple diffusion and radical transformation) are not common objectives in the policy fields under study.

1.1.3.2   Environment-oriented technology

Innovation is considered as a process, with technology being what is developed, modified, and applied through this process. Sustainable or environment-oriented innovation therefore concerns a particular class of technology, 'environment-oriented technology'. This is the same as 'environmental technology', defined by Gouldson and Murphy (1998: 29) as *"any technology that reduces the absolute or relative impact of a process or product on the environment."* We do not intend to draw a sharp distinction between environment-oriented and other technology, in either conceptual or practical senses. We use the term to indicate that the development and diffusion of certain kinds of technology is instrumental in the realisation of environmental policy goals. This means that it is a relative concept: (1) the environment-oriented nature of a technology is not necessarily intended by its developers; (2) the effect is largely dependent on the environmental properties of the conventional technology in use for the same purposes. In general, two different types of environment-oriented technology are distinguished: end-of-pipe technologies, and clean or integrated technologies. Again, we adopt the definitions used by Gouldson and Murphy (1998: 30):

> *"Control technologies are end-of-pipe additions to production processes which capture and/or treat waste emission in order to limit ultimate impact on the environment. Thus control technologies are distinct or stand-alone and commonly involve limited change to the process or product to which they are added. By contrast, clean technologies are general processes or products which fulfil a non-environmental objective as their primary purpose but which integrate environmental considerations into their design and/or application in order to anticipate and avoid or reduce their impact on the environment."*

At a given point in time, there is an initial stock of technological options (end-of-pipe and clean technologies) that are available to improve environmental quality, and this can be changed by a process of research and development (R&D) that results in a flow of new technologies that are added to the existing stock. Hence, there are always technologies which can be defined as the best available technologies (BATs) at any given time, but which can be further developed for future uses. The dynamics of these R&D processes need to be governed by the various actors to initiate and stimulate environmental innovation.

Environment-oriented technology may be classified under three headings: (1) pollution control technologies, (2) pollution prevention technologies, and (3) management systems (Klassen and Whybark, 1999):

1. Pollution control technologies only affect the production or manufacturing process at the end. This type of technology controls the environmentally harmful effects of an existing process, and includes both end-of-pipe controls and ex-post repair (which refers to cleaning up harmful effects of past practices, often driven by new regulations or by improved technical understandings).
2. Pollution prevention technologies involve fundamental changes to the basic products or primary processes to reduce negative environmental impacts. They include both product and process adaptations.
3. Management systems cover infrastructural and integrative investments that underpin both control and prevention technologies, or modify operating practices. In this context, the term 'technologies' is used in a general sense, e.g. employee-training for spill prevention, modified operating procedures, and the development of environmental measurement systems. Examples are compliance audits and monitoring systems (EMAS, ISO 14001, etc.).

1.1.3.3   Policy system/policy approach

Research on innovation-oriented environmental policy and its regulation mechanisms usually has a focus on specific policy instruments or tools and their impacts on innovation activities. The ENVINNO national studies presented in this book are based on a different research strategy, which defines environmental policy and environment-oriented technology policy approaches in a systemic manner. Policy instruments are elements of these approaches, but they are only one part of the regulation impact chain that consists of *"the institutional context, the actor constellation and policy learning in networks and negotiation systems"* (Jänicke, 1996). Therefore, an analysis of policy approaches needs to cover all these elements. For the specific ENVINNO research focus on an environment-oriented, innovation friendly policy sphere in particular, both environmental and environment-oriented technology policy approaches need to be analysed in this manner.

1.1.3.4   Policy integration and inter-policy co-ordination

The concept of inter-policy co-ordination of environmental policy and environment-oriented technology policy refers to the extent to which policy incentives from different policy fields are reinforcing one another. As this book is about public policy aimed at stimulating environment-oriented innovations, inter-policy co-ordination refers to the way both policy systems are supporting this particular aim. The first stage of inter-policy co-operation involves deliberate co-ordination of the employment of means, such as resources and instruments. The second stage involves joint initiatives by policy makers, such as joint programmes. The final step could be complete policy integration, when the boundaries between the EP and TP systems would vanish and one integrated policy system would occur. This stage however is not looming in any of the six member states, so it concerns a theoretical option and is beyond discussion here.

1.2   The EU policy context

The focus of this book is on the national level: it is about the way sustainable innovations at the company level are stimulated through environmental and technology policies in six EU member states. To understand these processes properly, one has to view them within the context of relevant policy developments at the EU level. This section therefore reviews the developments in the environmental policy system of the EU. The present objectives do not require a comprehensive historical review, it is sufficient to outline the major development lines and to pay some attention to the most relevant issues in the interplay between environmental policy and environment-oriented technology policy. However, a coherent picture requires a well-considered perspective on the complex and often chaotic developments of EU environmental policy. In this review, we distinguish a single underlying theme and several 'story lines' of formal policy documents and decisions by EU institutions.

The main theme is the historical development of the approach taken by the EU towards environmental problems. In a nutshell, this history has three stages. Firstly, in the 1970s, environmental policy was reactive: environmental problems were treated as isolated issues, and priority was given to stopping further pollution and repairing damage caused in the past. Secondly, in the 1980s, the interdependency of environmental problems was acknowledged; and the focus shifted towards prevention, for instance through substituting integrated technology for end-of-pipe measures. Thirdly, in the 1990s, the integration of environmental policy with other policy sectors prevailed, initiated by the Brundtland report (WCED, 1987), the 1992 Earth Summit in Rio de Janeiro, and the confirmation of sustainable development as one of the constitutional cornerstones the European Union.

Using this overall guideline, the review addresses the following story lines of formal EU policy:
- Within the realm of EU environmental policy, there are the successive Environmental Action Programmes, from the 1st EAP (1973-1976) to the

6th EAP (2001-2010), and the advance of the framework directives. The leading actor in this respect is the former DG XI, now DG Environment.

- Environmental integration: the process of integrating environmental concerns into all relevant policy sectors at the EU level. This was initiated by the Amsterdam Treaty of 1997, with the European Council as the leading actor. This process is the most evident analogy to policy integration between environmental and technology policies at the national level.
- The EU's Sustainable Development Strategy: the formulation of a long-term approach for the EU's contribution to the Rio-process, with the Rio+10 World Summit on Sustainable Development in Johannesburg in 2002 as the most recent landmark. Here, the European Commission is the leading actor.

As these story lines are interconnected to a large extend, the overall story is best presented in a chronological order. The main criterion for dwelling on specific issues is their relevancy for EP and ETP inter-policy co-ordination, for which there are two main issues. Firstly, the focus on prevention of environmental damage requires integrated or sustainable technologies, which is in itself a reason for integrating environmental and technology policies. In addition, the acknowledgement of the crucial importance of sustainable development has given rise to a revision of the traditional objectives of economic, social, and environmental policy, and of the meaning of technological development and innovation. Both issues will be addressed further.

1.2.1 United Nations and sustainable development

In a way, developments at the United Nations set the stage for EU policy developments, just as the latter have been setting the stage for national policy developments. The United Nations has played an important role in spreading awareness that environmental concerns are tightly related to social and economic concerns, leading to the formulation of the concept of sustainable development. For the present review of EU policy developments, three major achievements at the United Nations level should be mentioned at the beginning of this review: (1) the Stockholm Declaration of 1972; (2) the Brundtland report of 1987; and (3) the Rio Earth Summit of 1992.

With hindsight, the year 1972 can be seen as a major turning point. The Club of Rome report 'Limits To Growth' (Meadows, 1972) was published, boosting the awareness of environmental issues, especially the finiteness of our natural resources. In that same year, the UN Conference on the Human Environment in Stockholm marked the breakthrough of environmental policy at the supranational level (Kronsell, 1997). It is worth noting that already at this early stage of supranational environmental policy, individual states were being urged to adopt an integrated and co-ordinated approach to environmental policy.[1] In the aftermath of this conference, the United Nations Environment Programme (UNEP) was founded, which continues to play an important role at the supranational level.

The World Commission on Environment and Development, presided over by the former Prime Minister of Norway, Gro Harlem Brundtland, was established by the United Nations. With its report, 'Our Common Future' (WCED, 1987), this commission created what is perhaps the most important landmark in the history of environmental policy. Not only did it evoke a sense of urgency all around the world, it also changed our concept of the environment. The essence of the concept of sustainable development has three components: (1) the interdependency of ecological, social, and economic issues; (2) the global dimension, or the interdependency of the north and the south; and (3) the rights of future generations. The report saw that the integration of policy fields, and the dissolution of a strict division of labour between ministries and other governmental agencies, would be one of the most important institutional issues in the 1990s.

The significance of the 1992 Earth Summit in Rio de Janeiro has been the acknowledgement that sustainable development requires an integrative approach to economic development that includes all the environmental and social issues. At this conference, the international community adopted 'Agenda 21', an ambitious global plan of action for sustainable development. In addition, all participating countries were committed to develop a 'National Agenda 21', and a framework for a 'Local Agenda 21' was offered. Rio also marked a peak in optimism about the possibility of solving all the major problems in the world by joint efforts of governments and business. At that time, the outcomes of the conference were backed by most of the major economic powers, including the United States. A major offshoot has been the Kyoto protocol, aimed at dealing with the global issue of climate change. Now, ten years on, the optimism has vanished as the implementation of the agreements has turned out to be very problematic.

1.2.2   Beginning of EU environmental policy

The first phase of EU environmental policy can roughly be delimited by two events. The start is marked by the general sense of urgency that arose in the early 1970s, reflected, for instance, in the 1972 UN Stockholm conference. Since then, environmental policy has developed as a distinct EU policy field.

---

[1]   Declaration of the United Nations Conference on the Human Environment (Stockholm 5-16 June 1972) principle 13.

The transformation to the next phase, when environmental integration became the main theme, is marked by the 1986 Single European Act.

In this first phase, EU environmental policy was based on three consecutive Environmental Action Programmes. The 1st EAP (1973-1976) (EC, 1973) confirmed the earlier acknowledgement of the environment as a separate policy field for the EEC by the then six member states (until 1972). The main objective was to provide a place in the European legal system for the environmental regulations and standards that were being developed in the member states:

> "... ensuring that the improvement of living conditions and ecological factors, which must now be considered as inseparable from the organization and promotion of human progress, be integrated in devising and implementing common policies" (EC, 1973).

In those days, environmental policy was considered to be instrumental in meeting the heightened targets for welfare, economic growth, and the free market. Discursively this was seen as necessary to combat pollution and nuisance, and for reaching "... a harmonious development of economic activities and a continuous and balanced expansion ..." (EC, 1973). A major driver was the motive of preventing the emergence of trade barriers due to divergent environmental standards in the various member states. Therefore, product regulations were given the highest priority in the selection of standards to be harmonised (Liefferink and Andersen, 1997:12).

The 1st EAP contained a preliminary outline of cross-sectoral integration mechanisms, notably in the form of environmental impact evaluations for decision-making in sectoral policy areas:

> "Effects on the environment should be taken into account at the earliest possible stage in all the technical planning and decision-making processes. The environment cannot be considered as external surroundings by which man is harassed and assailed; it must be considered as an essential factor in the organization and promotion of human progress. It is therefore necessary to evaluate the effects on the quality of life and on the natural environment of any measure that is adopted or contemplated at national or Community level and which is liable to affect these factors." (EC 1973)

The main issue in the 2nd EAP (1977-1981) (EC, 1977) was to update and continue projects and measures started during the 1st EAP (Barnes and Barnes, 1999: 35). The environment as a distinct policy field was developed further with an emphasis on issues such as improving the quality of life, protection of the natural environment, non-damaging use of land, environment and natural resources, increasing public awareness, and personal responsibility for environmental protection.

Integration of environmental concerns into other sectors and policy fields, which later became known as 'environmental integration', as well as the 'polluter-pays principle', were first raised as explicit policy goals in the 3rd EAP (1982-1986) (EC, 1983). *"Progress in this field has been extremely slow ... "* (Liefferink and Andersen, 1997: 12).

1.2.3   Single European Act

The second phase of EU environmental policy is dominated by the stepwise introduction of the principles of environmental integration. The major steps have been: the 1986 Single European Act, the 1992 Maastricht EU Treaty, and the 1997 Amsterdam EU Treaty.

In the Single European Act (signed in 1986, and taking force in 1987), in itself a review of the original Treaty of Rome, the role of environmental policy was acknowledged for the first time in the EU Treaty. Article 130r included the major principles of environmental policy - precaution, prevention, rectification at source, and polluter-pays - as well as the need for environmental integration:

> *"Community policy on the environment shall aim at a high level of protection taking into account the diversity of situations in the various regions of the Community. It shall be based on the precautionary principle and on the principles that preventative action should be taken, that environmental damage should as a priority be rectified at source and that the polluter should pay. Environmental protection requirements must be integrated into the definition and implementation of other Community policies." (Article 130r, par 2, subpar 1)*

Subsequently, the issue of environmental integration has been granted higher priority in each revision of the Treaty. Firstly, it was indicated as a substantive issue in the 1990 Declaration by the Heads of State and Government, which was implemented in the 1992 Maastricht EU Treaty. It was stressed that environmental protection requirements must be integrated into the definition and implementation of other Community policies in order to achieve sustainable growth. The main discursive approach for integration was to be a mutual co-dependency of environmental concern and prosperity. The long-term success of the internal market, and the economic and monetary union, was seen as dependant upon the core EU policies being truly sustainable. Thus, certain EU policy areas were targeted for environmentally integrative approaches: industry, energy, transport, agriculture, and regional development.

This integration principle was promoted to a constitutional cornerstone in the revision of the EU treaty established at the Amsterdam European Council in 1997.[2] Reflecting the core sentiment of the Brundtland report, sustainable development was added to the existing economic and social goals in Article 2 of the Treaty of European Union[3], which states the basic goals of the EU:

> *"promote economic and social progress and a high level of employment and to achieve balanced and sustainable development, in particular through the creation of an area without internal frontiers, through the strengthening of economic and social cohesion and through the establishment of economic and monetary union, ultimately including a single currency in accordance with the provisions of this Treaty."*

In Article 6, the 'integration principle' was emphasised:

> *"Environmental protection requirements must be integrated into the definition and implementation of the Community policies and activities referred to in Article 3, in particular with a view to promoting sustainable development."*

These changes have been a major impetus for changes in the EU's environmental policy, but also for developments at the national level (Barnes and Barnes, 1999: 55).

1.2.4   5th Environmental Action Programme

The recognition of the concept of sustainable development at the UN (Brundtland report) and at the overall EU level (Amsterdam Treaty) has had consequences for the next level down: the European Commission and the Environmental Action Programmes. The 4th EAP (1987-1992) did not involve a radical transformation (EC, 1987). It focussed on the compatibility of environmental protection and economic targets; issues such as economic development, job creation, and environmental protection were to be considered as mutually supportive rather than conflicting goals. A special emphasis was placed on the environmental impact of the single market (c.f. Barnes and Barnes, 1999: 38).

A radical change in the EU environmental policy came with the 5th EAP, called 'Towards sustainability' (1993-2000), which included a strategy for achieving sustainable development. It can be conceived of as the European Commission's response to the Rio Conference and Agenda 21. The 5th EAP confirmed the basic principles in Article 130r of the Treaty, it took a systematic and long-term approach, and it introduced the notion of shared

---

[2]   The Treaty of Amsterdam was signed on 2 October 1997 and actually concerned two treaties: (1) the Treaty of European Union (Official Journal C 340, 10.11.1997, pp. 145-172), and (2) the Treaty establishing the European Community (Official Journal C 340, 10.11.1997, pp. 173-308).

[3]   This article replaced Article 130r in the previous Treaty.

responsibility between government and social sectors. Five target sectors were designated: industry, agriculture, transport, energy, and tourism. The Commission sought to implement the fundamental objectives through a new generation of policy instruments with an emphasis on market-based and voluntary instruments (such as negotiated agreements) and on integrated framework directives.

The integration process was among the priorities (key mechanisms for implementation) of the 5th EAP which called for the *"... full integration of environmental and other relevant policies ... through the active participation of all the main actors in society (administrations, enterprises, general public)"* (EC, 1993). The functional differentiation of EU sectoral policy and its administration was not questioned, although environmental concerns were to be substantially integrated into the structural funds, R&D programmes, information polices, and indicators of the EU. The member states were asked to integrate environmental concerns, into spatial and urban planning, in economic fees and levies, in state aid subventions, and in education (EC, 1993).

The 5th EAP led to a number of integrated policy formulations, new environmental indicators and institutions, and new advisory units within existing bodies and sectors. An important institutional outcome was the establishment of three strategic evaluation and discussion groups. These groups were to become involved in the internal and public debates on the effectiveness of EU environmental policies.[4] The integrative approach to the environment and the economy has been reflected in several strategic documents such as a recommendation on employment and the environment, focussing on how to foster green technological innovation in the member states, issued by the Commission in the wake of the Amsterdam Treaty in 1997. The discourse on integration was oriented towards economic restructuring, the withdrawal of support for major energy and raw material consuming industries, and the promotion of innovation policies which would create jobs and benefit the environment. The objective was to escape from old technologies: using benchmarking with 'eco-job' profiles; improving R&D for new cleaner technologies; and internalising eco-costs in prices, and promoting eco-taxes and eco-education.

1.2.5   The Cardiff process

The Amsterdam revision of the EU Treaty marked the onset of a comprehensive implementation process for environmental integration, which

---

[4] General Consultative Forum on the Environment and Sustainable Development, The Implementation Network of National Authorities, and the Environmental Policy Review Group.

has been a major issue at subsequent European Councils. It is known as the 'Cardiff process' because the European Council meeting held in this city in June 1998 is considered to be the starting point. For this occasion, the European Commission had prepared a strategy paper on the integration process, as asked of them at the Luxembourg European Council of December 1997 (EC, 1998). In this strategy paper, which was called 'Partnership for integration', the European Council - consisting of the heads of state and government - was invited to encourage an implementation process to cover all the relevant policy fields. However, the European Council decided not to start the process across the whole range, but to confine the first round to three policy fields: Transport, Energy, and Agriculture. The organisation of the process reflects the policy process co-ordination between the major institutions of the European Union. Formally, the integration process is a joint effort of the European Council, the Commission, and Parliament: the 'partnership' in the title of the strategy paper. The respective 'Council formations' - consisting of the ministers of the member states - were given the lead by being invited to prepare strategic plans for their specific policy fields. At the Vienna European Council of December 1998, the call was extended to the fields of Development, the Internal Market, and Industry; and at the Cologne summit of June 1999, also to General Affairs, Economic and Financial Questions ('Ecofin'), and Fisheries.

The process was reviewed at the Helsinki European Council of December 1999. The Commission had prepared a working document for the occasion, called 'From Cardiff to Helsinki and beyond' (EC, 1999b). Adopting it proved to be a burdensome process. Instead of a straightforward approach with a uniform environmental assessment methodology and agreements on specified environmental targets for sector policies related to overall environmental policy targets, the process was hampered by inconsistencies in concepts and indicators and, worst of all, in environment policy targets (Görlach et al., 1999; Schepelman, 2000). Overall, the Helsinki evaluation was not very profound:

> *"In Helsinki, however, the Council only stated that evaluation and monitoring must be undertaken so that the strategies could be adjusted and deepened, if necessary. The Commission and Council were called upon to develop adequate instruments and data." (Schepelmann, 2000: 11)*

Nevertheless, the Helsinki summit did demonstrate that sustainable development had become a major issue for the European Union. Two other major policy processes were set in motion. The Commission was asked to prepare a proposal for the 6th Environmental Action Programme, which was assumed to deal with medium-term objectives, as well as *"a proposal for a long-term strategy dovetailing policies for economically, socially and ecologically sustainable development"*, which was intended to be the European Union's contribution to the Rio+10 conference, the World Summit on Sustainable Development in Johannesburg in 2002. The Stockholm European Council of March 2001 decided to institutionalise the status of sustainable development by carrying out an annual review of the state of sustainable development in the European Union at each Spring European Council meeting.

As yet, the Cardiff process is not complete. According to the original schedule, all the sector strategies should have been available at the Göteborg European Summit of June 2001, but this was not realised. The Council was then invited to finalise and further develop sector strategies with a view to implementing them as soon as possible, and present the results of this work before the Barcelona Spring European Council in 2002. Further, the Commission was urged by the Council to co-ordinate the Cardiff process with the other major processes in formulating the European Unions mid- and long-term sustainable development strategy:

> *"The process of integration of environmental concerns in sectoral policies, launched by the European Council in Cardiff, must continue and provide an environmental input to the EU Sustainable Development strategy, similar to that given for the economic and social dimensions by the Broad Economic Policy Guidelines and the Employment Guidelines. The sectoral environmental integration strategies should be consistent with the specific objectives of EU Sustainable Development strategy."* (EC, 2001a)

Observers are critical of the results of the Cardiff Process. Although the process has obtained its own dynamics and will not easily be stopped, commitment by the individual Council formations charged with the development of an integration strategy is often lacking. Kraemer (2001) and Fergusson et al. (2001) note the inability or unwillingness to recognise the full range of environmental impacts of current policies and decisions, and to face the consequences of fundamental changes. Each Council has chosen a format of its own, and the development of integration strategies is not in line with the state-of-the-art, or suggestions made by the European Council, since they are often lacking elements such as clear targets and objectives, timetables and provisions for monitoring and reviewing. However:

> *"[c]omparatively good are the sectors with an obvious environmental relevance and which were already included as "target sectors" in the Fifth EC Environmental Action Programme (transport, industry, agriculture, and energy)."* (Kraemer, 2001: 4)

1.2.6   6th Environmental Action Programme

Although the 5th Environment Action Programme marked a substantial step towards environmental integration, the mid-term 'global assessment' was rather critical (EC, 1999a). It concluded that while progress was being made in cutting pollution levels in some areas, the achievement of the policy goals could be seriously obstructed unless breakthroughs could be established in areas as:
– the implementation of environmental legislation in the member states;
– further integration of the environment into economic and social policies;
– internalisation of the environment policy targets by social actors;
– new impetus for dealing with a number of serious and persistent environmental problems as well as emerging concerns.

The new 6th Environment Action Programme, called 'Environment 2010: Our Future, Our Choice', generally continues the lines of its predecessor, but may be characterised as having a more strategic approach (EC, 2001c). This is due to the fact that it adopts a longer time horizon (2001-2010), and that is constitutes the environmental component of the EU strategy for sustainable development. In particular, it sets out five strategic approaches that emphasise the need for more effective implementation and more innovative solutions:
– ensure the implementation of existing environmental legislation;
– integrate environmental concerns into all relevant policy areas;
– work closely with business and consumers to identify solutions;
– ensure better and more accessible information on the environment for citizens;
– develop a more environmentally conscious attitude towards land use.

Apart from the more strategic approach, the adjustments in 6th EAP also place an emphasis on four priority areas that have become more urgent:
– climate change,
– nature and biodiversity,
– environment and health,
– natural resources and waste.

1.2.7   Sustainable development as a cornerstone of EU policy

The issue of environmental integration has been developing along several lines. The renewal of the EU Treaty in 1997 was the starting point for: the integration of environmental concerns into other policy fields, the Cardiff process, and the further development of the Environmental Action

Programmes as the EU policy agenda for the environment. Another follow-up to the Treaty renewal has been the elaboration of the environment as one of the cornerstones of the EU. This has been reflected recently in two major issues: the revision of the EU strategy at the Lisbon Summit in March 2000, and the EU strategy for sustainable development, prepared for the 2002 World Summit in Johannesburg.

1.2.7.1   The Lisbon Process

The Lisbon Summit of March 2000 was seen as the onset of a revitalised overall EU strategy in response to the challenges of globalisation and the knowledge-driven economy (2000b). Competitiveness and the challenge of becoming the world's first 'knowledge economy' have been the major driving forces, but the type of economy aimed for was explicitly a sustainable one. The 'Lisbon process' was supposed to get input from three sides: economic, social (notably employment policy), and sustainable development. Its relevance for the present discussion is in its particular manifestation as environmental integration, and in the way that the EU sustainable development strategy is linked with its innovation policy.

Innovation has been a policy priority of the European Union and the member states for some time. Many policy plans have been drafted that relate to innovation in one way or another, and many measures and support schemes have been implemented. Their diversity reflects the diversity of the underlying conditions, cultural preferences, and political priorities in the member states, as will be demonstrated in the national studies in this book. A distinct EU innovation policy is of recent origin. It does not aim at formulating uniform standards and approaches that should be implemented through national policies in a top-down fashion, but rather a bottom-up approach that allows adaptation to national and regional specificities. The role for the EU level is essentially limited to facilitating, benchmarking, and the exchange of 'good practices' (EC, 2000a).

A major landmark in the development of this EU innovation policy was the First Innovation Action Plan of 1996 (EC, 1996), which contained the priority lines for action by the Member States. A second landmark has been the Communication by the Commission, called 'Innovation in a knowledge-driven economy', published in September 2000 (EC, 2000c). As a follow-up to the Lisbon Summit, it put forward broad policy lines and five priority objectives to enhance innovation in Europe. The main features of these objectives are:
– coordination and coherence of innovation policies across the European Union,
– reduction in regulations to lower innovation transaction costs,
– measures to stimulate innovative enterprises,

- facilitation of innovation networks,
- ensuring innovation is high on the agenda of all the actors involved, as well as of the general public.

Although the creation of favourable circumstances for innovations to be developed and diffused also supports sustainable innovation, it should be noted that the emphasis on sustainable development at the Lisbon Summit is almost entirely missing from this crucial document.

1.2.7.2   EU strategy for sustainable development
The EU's sustainable development strategy is seen as a commitment stemming from the Rio Declaration that was to be delivered at the 2002 World Summit on Sustainable Development in Johannesburg. A document was prepared by a team of Commissioners for 'growth, competitiveness, employment, and sustainable development', the so-called Prodi group, after its chairman the President of the European Commission (EC, 2001a). It has been written in the spirit of the Lisbon process, and the 6th EAP constituted the basis of the environmental component. According to the Commission, the following six issues pose the greatest challenges to sustainable development in Europe:
- combating poverty and social exclusion,
- dealing with the economic and social implications of an ageing society,
- limiting climate change and increasing the use of clean energy,
- addressing threats to public health,
- managing natural resources more responsibly,
- improving the transport system and land-use management.

1.2.8   Co-ordination of EP and ETP in EU policy

In this section, the recent developments in European environmental policy have been reviewed from the perspective of environmental integration. During the 1970s, environment has grown into a distinct policy field within the European Union and, as shown in the national studies that follow, similar developments have occurred in the Member States. The events at the EU level differ from those at the national level because of the central position - at least in a formal sense - given to sustainable development in the EU Treaty. The revision of the Treaty has been the start of a process, governed in a top-down way, of integrating environmental considerations into the most relevant of the other policy fields, through the Cardiff process. Although, from the perspective of the environment, this is a positive development, it does have a downside because the relative significance of the environment as a distinct policy field has been reduced, as became clear with the 6th EAP.

The intention behind environmental integration is sound, because environmental policy can only be effective if the impacts of almost all human behaviour are controlled, and this requires adaptations in many policy fields. At the EU level, this principle has been adopted to a much higher degree than in any individual member state, certainly than in those discussed in this book. An obvious way to explain this is to look at the different nature of policy at the two levels. Generally, the policy processes and their outcomes at the EU level are much further away from the people who are eventually affected. Not only are EU policy processes less visible to citizens than national policy processes, but also most of the outcomes need to be implemented through the adaptation of national legislation, and this takes a considerable amount of time. Even under these circumstances, with relatively little attention from the media and the public, it has been very difficult, in the case of environmental integration, to arrive at a common approach in all policy fields for performing environmental impact assessments and establishing policy targets.

Developments at the national level clearly show different patterns. In most cases, the acknowledgement of a need for inter-policy co-ordination in order to achieve environmental policy targets has led to the close involvement of other ministries (Economic Affairs, Energy, Transport, Agriculture) in environmental policy, for instance in the drafting of the national environmental plans that have been established in many Member States. Integration of environmental policy targets into other policy fields, analogous to the Cardiff process, has occurred to a much lesser extent. There is perhaps one exception, and that is technology policy, which takes us back to the subject of this book. As will be shown by the national studies, the field of technology policy is often supportive of other policy fields, and in several member states it has been linked to environmental policy, in particular in terms of stimulating sustainable innovations by industry. In these cases, environmental integration is not aimed at curbing the negative environmental impacts of sectors such as industry, transport, and agriculture; but rather at a positive contribution by tuning particular elements of technology policy to environmental policy targets. The six national studies that follow all contrast sharply with the events at the EU level. Most of the national policies discussed involve incentives aimed at directly steering behaviour. These incentives stem from conventional instruments for environmental policy, such as legal regulation, market-related instruments, and communication; or from technology programmes, the usual instrument of technology policy. Inter-policy co-ordination between these fields has been achieved by various means, with different kinds of barriers needing to be overcome. A major difference to the EU level is the smaller scale of the national policy processes, and the consequent larger role for bottom-up initiatives.

To sum up, the developments at the EU level not only constitute a relevant context for the national studies, these studies also contain specific experiences with issues of inter-policy co-ordination and environmental integration that may be valuable for the future implementation, at the national level, of the outcomes of the environmental integration process at the EU level.

1.3 Overview of the book

Next the six country studies are presented in alphabetical order: Austria, Denmark, Germany, the Netherlands, Spain, and the United Kingdom. These studies are elaborated versions of the 'macro reports' for the ENVINNO project, involving an analysis of the two policy fields, i.e. environmental and technology policy. The chapters are structured to reflect the development of both policy fields during the last three decades, their institutional framework, the most significant instruments implemented and finally the co-operation efforts (policy approach). Each presents an individual national analysis, based on the specific institutional and general policy framework. Therefore, the final chapter, called 'Synthesis' is not a comparative study, but an attempt to synthesise elements of six national policy in order to shed light on a European framework of environment-oriented innovation stimulation. It focuses on the policy approaches and the steps towards policy integration as trends and findings coming out of the six national studies.

REFERENCES

Andersen, M.S. and Liefferink, D. (eds.) (1997) *European environmental policy. The pinoneers. Issues in environmental policies.* Manchester and New York: Manchester University Press.

Barnes, P.M. and Barnes, I.G. (1999) *Environmental policy in the European Union.* Cheltenham and Northhampton: Edward Elgar.

Brunner, D. and Mellitzer, J. (1996) *Sunrise über Österreichs Betrieben. Die Verbreitung thermischer Solaranlagen in Unternehmen des Gewerbe und des Tourismus in Österreich. Eine empirische Bestandsanalyse,* Vienna: Thesis paper Vienna University of Economics and Business Administration.

Cooke, Ph., Boekholt, P. and Tödtling, F. (2000) *The Governance of Innovation in Europe. Regional Perspectives on Global Competitiveness.* London and New York: Pinter.

Dosi, G., Freeman, C., Nelson, R., Silverberg, G., Soete, L. (eds.) (1988) *Technical Change and Economic Theory.* London: Pinter Publishers.

EC (1973) Declaration of the Council of the European Communities and of the representatives of the Governments of the Member States meeting in the Council of 22 November 1973 on the programme of action of the European Communities on the environment. *Official Journal of the European Communities*, C 112, 20.12.1973. [1st EAP]

EC (1977) Environmental Action Programme 1977-1981. *Official Journal of the European Communities*, C 139, 16.6.1977. [2nd EAP]

EC (1983) Environmental Action Programme 1982-1986. *Official Journal of the European Communities*, C 146, 17.2.1983. [3rd EAP]

EC (1987) Environmental Action Programme 1987-1992. *Official Journal of the European Communities*, 328, 17.12.1987. [4th EAP]

EC (1993) Towards Sustainability. A European Community programme of policy and action in relation to the environment and sustainable development. *Official Journal of the European Communities*, No C 138/5, 17.5.1993. [5th EAP]

EC (1996) *First Action Plan for Innovation in Europe*, COM(96) 589 final.

EC (1998) *Commission communication to the European Council. Partnership for Integration - A Strategy for integrating Environment into European Union Policies*, COM (98) 333.

EC (1999a) *Global assessment. Europe's environment: what directions for the future?* COM (1999) 543.

EC (1999b) *Commission working document. From Cardiff to Helsinki and beyond. Report to the European Council on integrating environmental concerns and sustainable development into Community policies.* Brussels, 24.11.1999. SEC (1999) 1941 final.

EC (2000a) *Commission staff working paper. Trends in European innovation policy and the climate for innovation in the Union*, SEC (2000) 1564.

EC (2000b) *Presidency Conclusions*, Lisbon European Council, 23 and 24 March 2000.

EC (2000c) *Communication from the Commission to the Council and the European Parliament. Innovation in a knowledge-driven economy.* Brussels, 20 September 2000, COM (2000) 567 final.

EC (2001a) *Communication from the Commission. A Sustainable Europe for a Better World. A European Union Strategy for Sustainable Development. Commission's proposal to the Gothenburg European Council.* Brussels, 15.5.2001. COM (2001) 264 final.

EC (2001b) *Presidency Conclusions*, Göteborg European Council 15 and 16 June 2001. SN 200/1/01 REV 1.

EC (2001c) Environment 2010: Our Future, Our Choice. The Sixth Environment Action Programme of the European Community. *Official Journal of the European Communities*, No L 242 10.9.2002. [6th EAP].

Feldman, M.P. (2000) Location and Innovation: The New Economic Geography of Innovation, Spillovers, and Agglomeration. In: G.L. Clark, M.P. Feldman and M.S Gertler (eds.) *The Oxford Handbook of Economic Geography*. Oxford: Oxford University Press.

Fergusson, M., Coffey, C., Wilkinson, D., Baldock, D. (2001) *The effectiveness of EU Council integration strategies and options for carrying forward the 'Cardiff' Process.* London: Institute for European Environmental Policy.

Freemann, C. (1982) *The Economics of Industrial Innovation*. Cambridge (Mass): MIT-Press.

Görlach, B., Hinterberger, F., Schepelmann, P. (1999) *From Vienna to Helsinki. Environmental requirements in the process of integrating environmental issues into other policies areas of the European Union.* Study commissioned by the Austrian Federal Ministry of Environment, Youth and Familiy Affairs. Wuppertal: Wuppertal Institute for Climate, Environment and Energy.

Gouldson, A. and Murphy, J. (1998) *Regulatory realities. The implementation and impact of industrial environmental regulation.* London: Earthscan Publications.

Jänicke, M. (ed.) (1996) *Umweltpolitik der Industrieländer. Entwicklung - Bilanz - Erfolgsbedingungen*. Berlin: edition sigma.

Klassen, R.D., Whybark, D.C. (1999) Environmental management in operations: The selection of environmental technologies. *Decision Scienes* 3 (3): 601-631.

Kline, S.J. and Rosenberg, N. (1986) An Overview of Innovation. In: R. Landau, N. Rosenberg (eds.) *The Positive Sum Strategy*. Washington: National Academy Press.

Kraemer, R.A. (2001) *Results of the "Cardiff-Processes". Assessing the State of Development and Charting the Way Ahead*. Report to the German Federal Environmental Agency and the German Federal Ministry for the Environment, Nature Conservation and Nuclear Safety. Berlin: Ecologic.

Kronsell, A. (1997) Policy innovation in the garbage can. The EU's fifth Environmental Action Programme. In: D. Liefferink and M.S. Andersen (eds.) *The innovation of EU environmental policy*. Oslo: Scandinavian Academic Press, pp: 111-132.

Liefferink, D. and Andersen, M.S. (1997) The innovation of EU environmental policy. In: D. Liefferink and M.S. Andersen (eds.) *The innovation of EU environmental policy*. Oslo: Scandinavian Academic Press, pp: 9-37.

Meadows, D. (1972) The limits to growth. New York: Universe Books.

Murphy, J., Gouldson, A. (2000) Environmental policy and industrial innovation: integrating environment and economy through ecological modernisation. *Geoforum*, 31: 33-44.

Meffert, H. (1992) *Marketingforschung und Käuferverhalten*. Wiesbaden.

Nelson, R.R. (1993) *National innovation systems: A comparative analysis*. New York: Oxford University Press.

Schepelmann, P. (2000) *From Helsinki to Gothenburg. Evaluation of environmental integration in the European Union*. Study commissioned by the Federal Ministry of Agriculture and Forestry, Environment and Water Management. Vienna: Sustainable Europe Research Institute.

Schrama, G.J.I. (1991) *Keuzevrijheid in organisatievormen. Strategische keuzes rond organisatiestructuur en informatietechnologie bij het invoeren van een personeelsinformatiesysteem bij grote gemeenten*. Enschede, PhD dissertation University of Twente.

Schrama, G.J.I. (1998) Theoretical Framework. In: G.J.I. Schrama (ed.) *Drinking water supply and agricultural pollution. Preventive action by the water supply sector in the European Union and the United States*. Dordrecht: Kluwer Academic Publishers, pp: 19-42.

Schmalen, H. and Pechtl, H. (1992) *Technische Neuerungen in Kleinbetrieben: Eine empirische Untersuchung zur Einführung von elektronischer Datenverarbeitung in Handwerksbetrieben*. Stuttgart.

WCED (World Commission on Environment and Development) (1987) *Our common future*. Oxford: Oxford University Press.

Chapter 2

Environmental Policy and Environment-oriented Technology Policy in Austria

JUDITH KÖCK, UWE SCHUBERT AND SABINE SEDLACEK
Department of Environmental Economics and Management, Vienna University of Economics and Business Administration, Austria

2.1   Introduction

Environmental policy (EP) was in its original form mainly event-driven and often ad hoc. Urgent problems needed to be solved immediately without applying a more systematic approach. The Austrian environmental policy was rather strict and effective compared to other European countries. The mainly command-and-control oriented policy style forced companies to originally invest in end-of-pipe technologies and currently in cleaner technologies. Therefore, technology policy provided an enterprise support system with the aim of funding new technologies.

It is widely recognised that environmental policy and environment-oriented technology policy influence a company's behaviour in general. Do these policy fields also influence its innovative behaviour? In other words, is there a strong linkage between policy stimuli and companies? This chapter characterises the policy stimuli without analysing their specific impact on a company's behaviour.

The chapter starts with the analysis of both policy systems focussing on their general orientation and additionally on their potential linkage. For identifying new challenges in the current environmental policy and environment-oriented technology policy systems the historical background has to be pointed out briefly. The characteristics of the Austrian environmental policy and environment-oriented technology policy regime

25

Geerten J.I. Schrama and Sabine Sedlacek (eds.) Environmental and Technology Policy in Europe.
Technological innovation and policy integration, 25-58. © 2003 Kluwer Academic Publishers. Printed in
the Netherlands.

- its strengths and weaknesses – need to be specified for defining elements of future integration between environmental policy and environment-oriented technology policy which could guarantee more potential influence on a company's environmental performance. Possible strategies for the future integration of both policy fields are worked out in section 2.4.

The analysis reported is based on interviews with experts in the environmental and technology policy as well as decision makers in this field and already existing literature evaluating these policies.

2.2   Environmental Policy (EP)

2.2.1   The standard EP system in Austria

2.2.1.1   History
The Austrian environmental policy has been relying primarily on direct regulation, but – mainly in recent years – several approaches to meet new challenges have come up. To understand this course of development the more recent historical background needs to be briefly outlined.

Environmental policy has actually existed much longer than the word 'environmental protection'. Officially, however, it existed since the 1970s (the foundation of the Federal Ministry of Environment, BMU). The heydays of environmental policy were the 1970s and 1980s; remedying the basic problems of air and water pollution, and to a much lesser degree of soil contamination, were the most significant activities then. Most of the regulations adopted in that period were really following the German legislation in this field, to some extent Swiss models and in other areas, such as toxic chemicals, OECD recommendations were followed.

> *"As in most industrial countries, environmental regulation in Austria has a long history, if only with regard to specific problems. Measures for the protection of forests (against excessive felling of trees, against damaging smoke) can be found centuries back. A clause of the Civil Code – enacted in 1811 – contains a paragraph against pollution from nearby installations." (Lauber, 1997b).*

Figure 2.1, provides an overview of the recent environmental policy development in Austria beginning with the early 1970s. It shows the most important events and issues and the measures taken in reaction along a timescale.

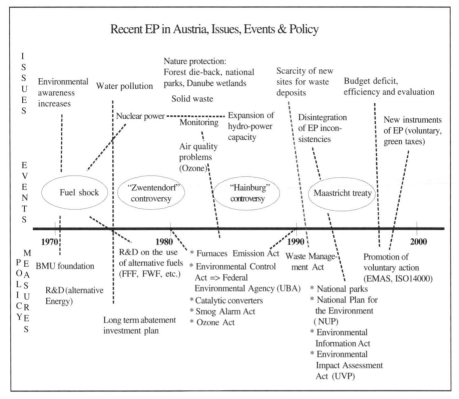

Figure 2 1: Recent environmental policy in Austria, Issues, Events and Policy Measures

The foundation of the 'Federal Ministry of Public Health and Environmental Protection' (BMU) in 1972 was more or less only of symbolic character. The ministry remained practically without responsibilities and opportunities for enforcement and almost without personnel for the task of environmental protection for about 15 years (Lauber, 1997b).

The confrontation concerning the nuclear power plant Zwentendorf about to start operation regarded as the first important environmental issue in Austria, led to a referendum in 1978 in which the population voted against the opening of the plant. As a result nuclear power generation was subsequently banned by law and the environmental protection movement gained in significance.

At the beginning of the 1980s the increasing awareness of environmental problems – not least in view of forest die-back – found expression in a rather emission-oriented legislation (e.g. Furnaces Emission Act 1980). For the first time the Austrian environmental law prescribed the application of the 'best available technology' and adopted the precautionary principle. In the mid-1980s legislation additionally concentrated on product and input-oriented instruments. An example for this assertion is the 'Amendment to the Motor Traffic Act' (1987) that required all new cars to be equipped with catalytic converters. Ambient environmental quality-oriented instruments – mainly minimum quality standards – played only a less important role (e.g. Smog Alarm Act 1989, Ozone Act 1992), compared to emission standards the handling of which in practice is relatively difficult.

Again a confrontation concerning electricity generation led to a renewed ardent public discussion on environmental issues. This time, the controversy centred on a dam for a hydro-power plant at Hainburg (on the Danube), the economic usefulness of which was doubtful and which would have destroyed some of the Danube's last wetlands (Lauber, 1997b). The subsequent 'Konrad-Lorenz Referendum' (1984) postulated the existence of an individual constitutional right of the citizens of environmental quality. These wetlands now constitute the core of a national park east of Austria's capital Vienna. The Austrian government reacted by implementing a constitutional law that declared comprehensive environmental protection as one of the goals of national policy.

With the introduction of the Environmental Control Act in 1985 the Federal Environment Agency (Umweltbundesamt, UBA) was established to serve as an advisory board ('think-tank') and to monitor environmental data for the Federal Ministry of Environment, Youth and Family. The studies made by this agency form the basis for planning and implementing environmental policy measures, and also for handling parliamentary questions concerning criticisms on drafts of proposed laws and ordinances as well as questions about current environmental problems.

The discussion on the principle of prevention instead of repairing, prevalent in the 1980s, resulted in the introduction of the Waste Management Act (1990) that undoubtedly heralded a new approach. At about the same time a draft bill on environmental liability supporting the 'polluter-pays principle' was prepared, but the recession beginning in 1992 intensified the opposition by business, trade and industry.

The 1990s brought a new legislative approach focussing on the importance of information. Thereby the impact of the EU, which Austria prepared to join, was clearly observable. On the basis of the Environmental Information Act (1993) everyone has the right of information on environmentally relevant data. These include information on the quality of the environment (condition of water, air, etc.) as well as data concerning public and private projects as far as they affect the state of the environment. The Environmental Impact Assessment Act passed in the same year on the one hand prescribes an integrated permitting procedure for large-scale projects with a (potentially) high impact on the environment and on the other hand emphasises the participation of citizens (Glatz, 1995).

The EU accession undoubtedly had an important impact on the Austrian environmental policy, expressed mainly in the form of the necessary change in the legal framework. The main benefits of this development accrued primarily in the Federal Ministry of Environment, Youth and Family giving it more influence on the management of environmental change in companies (see EMAS[1]). This ministry is also now a formal partner in decision making in Brussels, the 'social partners' (interest lobbies) are represented in Brussels as well, but they are not directly involved in negotiations. The *Länder* (see below) as well had to accept a decline in their influence resulting from the transfer of policy agenda to Brussels.

The OECD report on environmental performance in Austria published in 1995 conveys a rather positive impression, stating that the direct regulation approach, end-of-pipe technologies and subsidies led to respectable results, such as the decrease of SO_2 emissions (Lauber, 1997a). Many of the new challenges, however, cannot be met this way, that is why in recent years several new approaches have been developed (see section 2.2.2).

2.2.1.2   The major institutions

Austria is a federal republic, it consists of nine states or *Länder*. The political system is generally characterised by a division of responsibilities between the federal, provincial and municipal level. The federal government has extensive environmental responsibilities (e.g. in the field of air quality, hazardous waste, permits for industrial installations, steam boilers and engines, water management, forests, mining, traffic, and legislation on environmental impact assessment) whereas the *Länder* most notably are in charge of nature protection, non-hazardous waste disposal, airborne emissions from heating systems, construction norms, and the carrying out of environmental impact assessments. The municipalities handle matters such as waste collection, some of them run public enterprises for public transportation, gas or electricity provision, and they are responsible for the management of land resources (via zoning and building regulations).

At the federal level three ministries have a rather strong importance regarding environmental issues. The Federal Ministry of Environment, Youth and Family (founded 1972 as Federal Ministry of Public Health and Environment) is in charge of general environmental policy, it has to share this task in many situations with other federal agencies, usually the Federal Ministry of Economic Affairs (environmental questions related to energy policy, mining, tourism and road construction) and the Federal Ministry of Agriculture and Forestry (water legislation, water and groundwater management). In the year 2000 the Federal Ministry of Agriculture and Forestry and the Federal Ministry of Environment, Youth and Family were

---

[1]   Environmental Management and Audit Scheme.

merged. The former agricultural ministry started as the strong partner after this merger which resulted in a significantly weaker position of environmental tasks. As the Austrian policy generally is very consensus-oriented the so-called 'social partners' (Chambers of Commerce, Labour and Agriculture, the Industrialists' Union, and the Austrian Trade Union Federation) are integrated in the early stages of the political process, not only to safeguard the interests of their members, but above all to avoid (public) conflict situations. Moreover, all political parties in Austria assert a commitment to sustainable development although their understanding of this concept varies greatly according to their other priorities. By establishing the Federal Environment Agency (UBA) in 1985 an institution was founded to monitor and collect environmentally relevant data, which are published regularly.

Between 1984 and 1993 most of the *Länder* established an office of environmental advocates *(Umweltanwaltschaft)* to take up environmental concerns in administrative proceedings governed by provincial law (Lauber, 1997b).

Like in most European countries the importance of NGOs and green interest groups has risen continuously. Many of these are now fairly large (inter)national organisations (such as Greenpeace, Global 2000, WWF)[2], represented at the capital as well as in Brussels, which exert a considerable influence on environmental policy. There are, however, still many grassroot organisations active at the local level, still the backcloth of the environmental movement in Austria, providing inputs to policy formation and control. These activities are strongly supported by neighbourhood and stakeholder rights, often granting legal possibilities for intervention.

2.2.1.3   Instruments

The Austrian environmental policy system mainly relies on 'command-and-control' measures, as well as subsidies. During the 1980s the Austrian legislation focused on abatement equipment and production processes mainly (emission orientation). Later on the environmental aspects of products were increasingly discussed and led to new legislation (input and product orientation). Over time a trend towards greater policy integration can be identified especially under the impression of the EU accession. In recent years voluntary instruments, above all voluntary agreements (mainly to promote recycling efforts by industry), were introduced. In this field the EU regulation on voluntary environmental auditing (EMAS) needs to be mentioned. Contrary to most of Austrian legislation EMAS does not prescribe technical standards to be followed, but counts on 'overachieving'

---

[2]   These organisations are co-ordinated by the Centre of Austrian Environmental Organisations *(Ökobüro).*

legally imposed standards, the effect of public information and the introduction of an environmental management system in companies to ensure continuous awareness and action.

Below the most important regulations and their principal orientation, constituting the core of environmental law in Austria, are listed:

Emission-oriented regulations:
– Clean Air Act for Boiler Plants (1988): based on the Furnaces Emissions Act (1980,. regarded as the first 'true' environmental law in Austria, introducing the precautionary principle and the 'BAT-standard';
– Amendment to the Motor Traffic Act (1987): required all new cars to be equipped with catalytic converters;
– Water Rights Act;
– Waste Management Act (1990).

Input-oriented/product-oriented regulations, such as:
– Several laws restricting the sulphur content of fuels;
– Chemicals Act (1987) (regulating the release of substances such as CFC, asbestos);
– Pesticide Act (1991).

Ambient environmental quality-oriented regulations, such as:
– Smog Alarm Act (1989);
– Ozone Act (1992).

Others, such as:
– Environmental Control Act (1985): establishing the Federal Environmental Agency (UBA) and obliging the Federal Ministry of Environment to monitor the condition of and changes in the environment continuously;
– Environmental Information Act (1993): obliging industry to provide the general public with environmentally relevant information;
– Environmental Impact Assessment Act (1993): integrating permitting procedures for large-scale projects with a (potentially) high impact on the environment.

Some elements of an incentive based policy system exist, such as:
– Energy Tax on natural gas and electricity (low financial incentive; additional, no reduction of other taxes);
– Tax on Fossil Fuels: earmarked for road construction and maintenance;
– Car Tax (NoVA): since 1993; sales tax rate increases progressively with average fuel consumption, thus provides an incentive for more energy efficiency (mostly smaller cars);

– Deposit-refund system (bottles, refrigerators).

2.2.1.4   *Technology-related strategies*
For years Austrian environmental policy has relied on the prescription of technical standards (best available technology – BAT), the installation of filters and repair activities. Examples are:
– With respect to Air Pollution Control: Obligatory installation of filters (very stringent emission standards).
– With respect to Wastewater Treatment: Obligatory purification (construction of abatement facilities).
– With respect to Contaminated Land: Clean up obligation.
– With respect to Waste Management: Separation of solid waste and recycling.

The policy system can generally be characterised by equipment-oriented emission standards. The major gaps of this approach are the exemption of private households and the lack of co-ordination of emission and ambient environmental quality standards.

Only in recent years the principle to prevent emissions instead of repairing has become the guideline of new regulation and policy formation. The Waste Management Act (1990) illustrates this, by primarily aiming at the avoidance of waste instead of waste utilisation or treatment. This strategy requires co-operation between the authorities and business enterprises, thus introducing co-operation as a major new feature of environmental policy in Austria.

With the compilation of the *National Plan for the Environment* (NUP) in 1995 a new long-term strategy was designed to provide an opportunity to initiate the necessary structural changes in a substantial, consistent and enduring manner. It is meant to serve as a guideline for policy, should integrate environmental aspects into all government policies and gives clear priority to cleaner technologies. The key elements of the plan include long-range quality targets based on scientific criteria as well as proposed measures to reduce pollutants, the sensible use of non-renewable resources and the minimisation of material flows. However, as this plan is not legally binding, substantial changes are not necessarily to be expected.

Increasingly, in search of instruments that are less costly to the public than the enforcement of formal command and control measures, voluntary agreements have been promoted. Most of them are signed by government and business to achieve specific recycling goals. They enable the parties concerned to take more responsibility and are rather flexible, but are exposed to criticism for not being published, thus it is difficult to check on their success or failure.

As far as the implementation of price-related economic incentives is concerned, although the taxation of natural gas and electricity was a first step, these instruments have not been adopted seriously yet. A 'real' energy tax seemed possible in 1995 but, like in many other European countries, was (again) delayed, the discussion is still ongoing.

Since the EU accession new concepts have been determined by an obvious restraint of decision makers. Austria has no longer the intention to figure as a presumed 'pioneer', many governmental and business organisations claim further progress should be achieved only in accordance with EU changes. However, there are almost no political alliances between the European environmental 'pioneers' (Austria, Denmark, Germany, Finland, Sweden and the Netherlands) and so it is harder to introduce new national strategies as well, lacking the pressure from Brussels.

2.2.2   New developments

2.2.2.1   National Plan for the Environment (NUP)
The National Plan for the Environment characterised by its long-run perspective which was adopted recently, is meant to serve as a binding guideline for future actions. It constitutes a chance to orient the necessary structural changes to be made in this country by environmental quality goals. Work on this plan was started in 1992. The purpose of seven working groups was to contribute to the operationalisation and implementation of the concept of sustainable development in Austria, particularly with respect to the policy areas relevant for environmental quality. This work included the definition of long-run strategic ecological aims of a qualitative as well as quantitative nature spanning the various environmental media and sectors of the economy. Additionally medium- as well as long-term integrative concepts of preventive environmental policy were to be formulated and to be propagated and well established in the public. Thus, a long-term concept was developed which operationalises the political will to achieve an integration of all environmental policy issues at all levels of policy making, particularly in industrial, transportation, energy, agricultural and public health policy as well as the future R&D and technology policy areas.

It is not the intention of this first environmental plan to prescribe a static list of policy foci instruments and measures. It is rather the goal to set in motion a dynamic process promoting not only planning activities and goal formulations but to contribute to the implementation of these plans as well as their evaluation. The latter is to take place regularly every two years after the adoption of plans and should result in amendments and further development of the National Plan (federal government declaration, 1995).[3]

---

[3]   This intention has, however, not been followed yet.

*2.2.2.2   Economic incentive based instruments and voluntary
             environmental improvement programmes*

These instruments intended to provide incentives for better environmental performance are not very common in Austria. One of the oldest instruments in this field is the deposit-refund system (bottles, refrigerators) that was introduced to meet specific recycling goals. The 'cost-sharing contribution' (*Altlastenbeitrag*) charged on the deposition, the intermediate storage and the export of waste can be mentioned in this field as well (Streißler, 1997).

In recent years the introduction of green taxes has been heavily discussed but still little decisive action has been taken. Some taxes considered to promote environmental goals today originally served other purposes, most of all the taxes on mineral oil and a sales tax on motor vehicles graded by energy efficiency. The introduction of a moderate tax on natural gas and electricity in 1996 does not necessarily qualify as a green tax because of its low financial incentive.

Summarising it can be stated that significant tax based incentive policies have not been introduced in Austria, yet.

Another relatively new approach are voluntary instruments such as voluntary agreements or the EU regulation on voluntary environmental auditing (EMAS, 1995). The 'Federal Environmental Agency' establishes, revises and updates the list of registered sites (currently about 170 in Austria) and the list of accredited environmental verifiers. Austria has made use of the possibility to extend the application of the EMAS regulation to services (Environmental Verifier and List of Sites Act – UGStVG (Federal Law Gazette No. 622/1995) valid for the transportation and banking sectors on an experimental basis). The interest in EMAS is relatively high in Austria, not least because of specific subsidies. Moreover, many business enterprises may be sensitised to environmental protection by the fact that larger firms are required to designate a person responsible for environmental questions. The fact that EMAS does not aim at any specified target seems rather problematic and therefore specific environmental and even less technological priorities cannot be supported very well this way.

Opinions as to the effectiveness of this policy approach differ widely. This is reflected in the two following statements (source: ENVINNO-interviews):

"If an eco-audit is carried out properly, I would bet that for example 100 paper factories, would choose a hundred different solutions to the emission problem, with better results than a command and control approach demanding uniform emission standards could ever achieve."

"EMAS, in my opinion, is not an effective policy instrument at present. As everybody knows, legal compliance is not checked properly, so certification is mostly not justified. This is just a marketing trick, the

quantitative success in terms of participants in the scheme is mainly due to the subsidies available."

Sectoral covenants are another instrument of a voluntary participation nature. The Federal Ministry of Economic Affairs has initiated and promoted such covenants in some economic sectors over the last couple of years, predominantly in the area of material recycling. Presently 16 such sectoral agreements exist (see list in the appendix), the functioning of which depends crucially on the market value of the materials to be recycled. (Report of the Federal Ministry of Economic Affairs, 97/98).

2.2.2.3   A new emphasis on clean technology

Until the end of the 1980s environmental protection was almost exclusively considered synonymous with the use of specific technological solutions to reduce emissions. End-of-pipe technologies applied by industry were to reduce emissions at the production site considering only the environmental medium into which emissions were discharged. This media-specific approach makes a comprehensive view more difficult and obviates integrative policy efforts. Frequently it simply leads to shifts of waste loads from one medium to another or from one region to another (Kanatschnig, 1986). The increasing global pollution demonstrates that this type of policy cannot guarantee success in the long run.

The efforts made in the 1980s to promote R&D in environment-related technology (mostly end-of-pipe) with the hope of developing a pioneering role which could open up export markets did not materialise to the degree expected. This is mainly due to the fact that many countries did not choose to embark on major clean-up programmes, thus not offering chances to sell these technologies (this problem was first experienced by the Japanese industry and later, in the 1990s Europe followed). The shattered hopes for new, large markets and the economic recession in the 1990s with its concomitant shifts in public concerns and priorities thus led to a general slowdown in environment related investment and technological development.

As the media-focussed policy had run its course and that the end-of-pipe technological thrust had lost its momentum a new policy approach was warranted. This realisation led to new concepts attempting to reach a higher degree of integration between environmental policy and TP (see the following sections).

2.2.3   Evaluation of Austrian environmental policy

The interviews conducted with experts as well as governmental evaluation documents yield the following list of the most significant strengths and weaknesses of Austrian environmental policy:

Strengths:
– Environmental awareness in Austria is high.
– The most serious problems have been solved.
– Policy formation is consensus-oriented, discussions are open.
– Stringent emission standards.
– A tradition of environmental policy exists now, with more or less clear responsibilities and tasks.

Weaknesses:
– Lack of safeguards against environmental risks of further economic growth.
– Problem anticipation is underdeveloped.
– The legal framework assigning tasks and responsibilities to the various tiers of government is not transparent, sometimes international and focussed enough.
– The density of regulations is so high now that transparency is lost and the control of implementation has become too (public) resource-intensive, thus often sporadic and ineffective.
– Austria has no longer the intention to figure as a presumed 'pioneer', further progress should be achieved only in accordance with EU changes.
– Environmental issues are given lower priority recently.

The system of regulating emissions at the level of individual production units, which has been successful in Austria can cope with continued economic growth only to a certain degree. Once all production processes are in compliance with the law and the concomitant 'best practice' techniques are applied everywhere, increases in the volume of production are necessarily accompanied by larger flows of material and energy and, given the technology, must lead to higher emissions. (Such processes can be observed, e.g. with catalytic converters in California.)

Complying with ambient environmental quality standards in the long run makes a limiting of aggregated emissions in all areas of a country mandatory hence imposing 'macro-emission quota' is not avoidable in the long run. This perspective has become particularly evident in the debate about climatic change and CO_2 emissions on a global scale. A process of air quality management based on quotas is a necessary consequence, a hard political task to solve, obviously. It becomes clear that the standard command and control approach has come under fire, other new instruments might be better equipped to cope with this challenge.

2.3   Environment-oriented technology policy in Austria

Environment-oriented technology policy is part of the general Technology Policy (TP) of a country. It can offer incentives for companies to innovate, influence the type of innovation and convey to the users of such policy programmes. Alternatively funds can be earmarked for specific activities, such as R&D and innovation with the goal to reduce pollution. Both systems exist in Austria side by side, targeted programmes being a more recent development. In the following paragraphs the elements of both approaches are briefly outlined.

Figure 2.2 on the next page provides (similar to figure 2.1 concerning environmental policy) an overview of the most important events and issues with respect to environment-oriented technology policy development in Austria, starting with the late 1960s.

2.3.1   The evolution of technology policy since the Second World War

Contrary to some of the industrial countries in the 1950s and 1960s, an explicit technology policy did not exist in Austria and R&D efforts by industry were generally negligible. Much of the necessary technology was imported from abroad. Considerable economic growth resulted from the import of capital goods from the US supported by the Marshall Plan (the European Recovery Programme). However, in the long-run this 'imitation' strategy caused several problems (e.g. brain drain of Austrian scientists, clinging to second-rate technology, increasing conditions of dependence) and finally led to some rethinking. Furthermore, several studies stressed the importance of technological innovation, particularly the support of independent technology development. This recognition came in most industrialised countries in the 1970s and early 1980s, thus leading to a development of an innovation policy replacing and complementing the traditional science and technology approach. In Austria this phase was initiated only in the late 1980s. It was also in this period that the social, political and ecological impacts of innovation entered the stage of public interest and the entire context of these processes became an important issue (see Gottweis and Latzer, 1997).

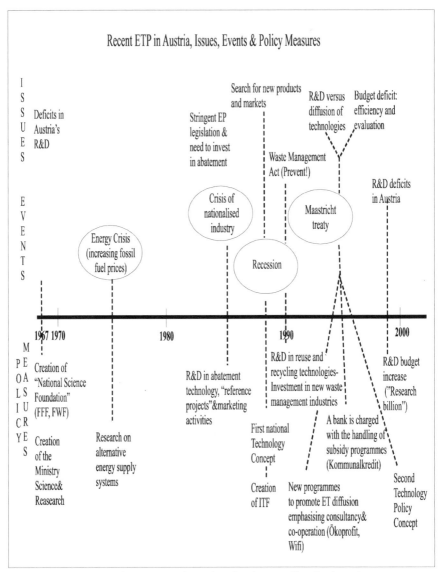

Figure 2.2: Recent environment-oriented technology policy in Austria

The recession plagueing the industrialised world in the late 1980s and beginning 1990s, evidently expressed in the highest unemployment rates since the war in some countries and very slow productivity growth led to intensive analysis of the phenomenon of technological change and more explicit policy programmes (Hofer et al., 1998). In Austria this trend was also accompanied by the problems in the nationalised industries, mostly active in the basic materials sector, which had been the engines of early economic growth after the war. The belief that only large industries in the country needed continuous investment support and increasingly subsidies from the tax payers to retain jobs, led to a general neglect of science and technology as well as innovation policy in favour of investment subsidies as an employment promotion strategy (see Aiginger, 1997). The crisis of the nationalised industry cumulated in an organisational reform of this industry which eventually led to a complete privatisation. The loss of markets (basic materials, steel, etc.) led to a search for new and international market niches. Environmental technology was seen as one of the chances to develop new product lines.

The implementation of the Research Funding Act in 1967 provided the legal framework and the actual starting point for a warranted explicit technology policy system. This act formed the basis for the foundation of two research promotion funds still in action today, i.e. the 'Austrian Industrial Research Promotion Fund' (FFF) and the 'Austrian Scientific Research Promotion Fund' (FWF) as well as for the establishment of a Federal Ministry of Science and Research (today's Federal Ministry of Science and Transport, BMWV) in 1970. Two years later this ministry presented Austria's first 'Research Concept'. Subsequently the research expenses and the total volume of research personnel increased considerably. With respect to environment-related issues, within the established framework a focus of R&D on energy supply alternatives evolved.

Environmental policy triggered technological developments to comply with regulations (e.g. furnaces emission act). These new techniques proved to be the market chance looked after. Further R&D was needed to maintain and expand the share in this new 'environmental product' market. As a consequence specific technology programmes were needed. This marked the beginning of an earmarked TP oriented by the environment (Köppl and Pichl, 1995).

The first 'Technology Policy Concept' in 1989 constitutes a cornerstone in this policy field in Austria. Its most important features are:

Concerning the aims:
- Improvement of co-operation and the information and knowledge transfer between universities and enterprises in Austria and abroad.
- Promotion of R&D by providing the necessary financial and legal basis for these activities (e.g. by introducing appropriate environmental regulation) to stimulate demand and to facilitate the necessary marketing activities by producers of new products and services.

Some of the measures foreseen were:

- Better planning and co-ordination by and between different parts of the public administration as well as stimulation of regional and local initiatives in this field.
- Provision of new educational opportunities in technologically relevant fields as well as management training possibilities to facilitate innovation processes at the micro-level (e.g. by setting up university extension programmes, emphasising and supporting technology-oriented university curricula, etc.).
- Support for newly founded enterprises based on new technology development.
- Promotion and sponsoring of international co-operation (e.g. participation of Austrian researchers in EU programmes).
- Purchasing policy of governmental organisations.
- International orientation, following developments abroad, became an explicit goal of policy efforts.

Among the programmes to promote innovation several initiatives to increase environmental awareness and information and demand for appropriate technologies need to be mentioned in this context. A targeted policy focus on 'Environmental Technology' existed between 1988 and 1995. The objective of this programme was to stimulate environment-related innovation in industry, not only for environmental policy motives, but also to increase the competitiveness of Austrian companies. Clean technologies were now clearly favoured and prevention declared to be preferable to emission control.

At the beginning of the 1990s waste management became a dominant policy issue (existing dumping sites were close to full capacity, new sites politically close to impossible to identify and arrange). Research activities (and expenditures) in this problem area increased considerably, a focus on reuse and recycling and evolution of a waste management industry was established.

New strategies in research and technology policy in recent years have led to reforms in research, teaching and training, to a number of measures for deregulation and decentralisation and to more self-administration and individual responsibility of the implementing institutions. Due to the dominant position of the universities in Austrian science changes in university strategies have a great impact on the technology policy as well. So the 'University Organisation Act' (UOG) of 1993 that assigned Austrian universities a wider scope of autonomy with regard to staff and financial matters considerably will offend the technology policy system in the future. The latest 'University Organisation Act' (UOG) of 2000 will release universities with full autonomy.

2.3.2   The present state of environment-oriented technology policy in Austria

2.3.2.1   New technology policy concepts

The first time an explicit effort to introduce a specific environment-oriented technology policy was made, was in 1994 (draft of a Technology Concept). It formed the basis for the 'Second Technology Policy Concept' in 1996 (which also introduced a new focus on social demands). In this document the federal government acknowledges explicitly the need for an institutionalised dialogue between technology and environmental policy agents.

The figures presently show that all the predicted trends (furthered by the new policies) have become a reality also in Austria. The products and production in general is becoming ever more knowledge-intensive, sources of knowledge are utilised increasingly and the actors involved are becoming more numerous and more diversified. The creation of new knowledge via R&D is recognised as one of the most significant policy challenges facing the country. Despite this fact many shortcomings do still exist. Prominent among them is the lack of a continuous allocation of funds for science, R&D and innovation.

Principle aims of the Technology Concept 1996 are the improvement of the competitiveness of the economy and of social and ecological conditions. The following key strategies were listed in this connection:
- *Enhancement of research capacities*: Support of knowledge-based enterprises by combining strengths of Austrian industry and Austrian science. Promotion of high-quality education and highly efficient research facilities. Establishment of polytechnics as well as non-university based research institutes and special R&D organisations under contract for government research.
- *Promotion of new technology adoption*: Establishment and support for technology transfer centres and high tech parks.
- *Improvement of material infrastructure and internationalisation*: Support of material infrastructure (transportation, telecommunication, energy supply, etc.) and incorporation into a European net of infrastructure.

In addition to economic competitiveness and social goals the present Austrian technology policy explicitly aims at the achievement of environmental objectives. The following technological innovation strategies have been defined as being of prime importance:
- Clean technologies instead of end-of-pipe technologies;
- Minimisation of material and energy flows;
- Avoidance of toxic substances (see figure 2.3).

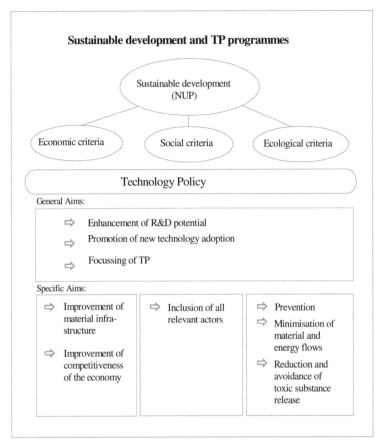

Figure 2.3: Sustainable development and TP programmes

A special programme to promote Sustainable Economic Activities was introduced in the framework of the new Technology Concept. The main objective of this specific environment-oriented technology policy is to promote structural changes in the Austrian economy in the direction of increasing eco-efficiency. The warranted R&D activities as well as the expedient adoption of new environment-oriented technologies are financed out of an earmarked fund ("Austrian technology billion") financed out of the federal government's budget. The amount of funds available for the environment-oriented activities is €2.9-3.3 million.

In the above-mentioned programme, in accordance with the National Plan for the Environment (NUP), the following policy foci were established:
- Improvement of energy efficiency and increased utilisation of renewable sources of energy.
- Promotion of renewable basic raw materials.
- Development of new products, services and new production processes.

- Support of regional concepts and efforts towards achieving sustainable development.
- Providing incentives for institutional and structural innovation.

2.3.2.2 The major institutions

A formal distribution of responsibilities concerning the Austrian TP does not exist, in the Austrian 'Federal Ministry Act' (Bundesministeriengesetz) no specific regulation can be found. As far as federal government activities in this field are concerned the activities of two ministries have had a rather strong impact on TP issues, namely the 'Federal Ministry of Science and Transport' (BMWV) and the 'Federal Ministry of Economic Affairs' (BMwA). Today's Federal Ministry of Science and Transport is the result of a recent merger between the long-standing 'Federal Ministries of Science and Research' (BMWF, since 1970) and the 'Federal Ministry of Transport' (BMÖWV, which has been in existence since 1896). With regard to 'science' the BMWV is responsible for secondary education, the universities, government research and the promotion of research, as well as the promotion of technology for commercial development.

Due to the lack of an appropriate authority to co-ordinate the relevant measures and activities, the Federal Chancellery undertook the definition and development of an integrated research and technology policy concept (see above). At the federal level the 'Federal Ministry of Environment, Youth and Family' should also be mentioned in this field regarding environment-oriented technology policy. It also plays a role in 'environmental policy supporting schemes' implemented by the 'Austrian bank for loans to local administrations' (ÖKK) mentioned below.

Several research funding institutions that finance basic and application-oriented research and technology developments can be identified (including environment-oriented technology policy based R&D implicitly):

- *Austrian Industrial Research Promotion Fund* (Forschungsförderungs-fonds für die gewerbliche Wirtschaft, FFF): created by the Research Funding Act (FFG 1967). The task of this fund is to finance innovative projects in applied business-oriented research; support is given in the form of loans, interest rate subsidies and the assumption of liability.
- The total volume of the fund 1998/99 was €138 million.
- *Austrian Fund for the Promotion of Scientific Research* (Fonds zur Förderung der Wissenschaftlichen Forschung, FWF): created by the Research Funding Act (FFG 1967) as well. It was set up as an instrument for general and targeted research funding and is supervised by the BMWV. Its main emphasis is on individual project financing. Lately special, targeted research areas have also been supported by this fund.
- The total volume of the fund in 1998 was app. €58 million.

- *ERP Fund* (European Recovery Programme): It focuses on increasing economic growth in an international context and the application of innovative technologies. The fund has a sum of €2.18 billion from the former Marshall Plan at its disposal. It mainly operates via direct research funding and interest-supported loans.
- *Innovation and Technology Fund* (Innovations- und Technologiefonds, ITF): It was installed by the Federal Government (ITF Act (1987), regulating the Innovation and Technology Fund). It supports national and international R&D projects. In 1988 the ITF started a first special targeted programme on 'environmental technologies' that is primarily focussed on production activities, specifically on:
  - the support of clean technologies;
  - increasing the share of 'green' products and consumer awareness;
  - the propagation of EMAS and ISO 14000 certification. The total volume of the fund in 1998 was app. €29 million.

The BMwA and the BMWV are responsible for the activities, the board of trustees is headed by the Federal Chancellor. In addition other ministries, interest groups as well as the two largest parties represented in Parliament and last, not least other institutions promoting technology participate in the activities of the ITF. Additional tasks of this institution are its functions serving as a platform for discussion and co-ordination:

- *BÜRGES*: supervised by BMwA (founded 1954), primarily targeted for the funding of the activities of small- and medium-sized enterprises, by guarantees for bank loans, equity capital and internationalisation, interest rate caps, premia and subsidies.
- *Innovation Agency*: The Agency, established in 1984, is a non-profit organisation whose business is the support and encouragement of ideas and assistance in implementing innovation projects.
- *Association of Austrian Technology Centres* (Vereinigung der Technologiezentren Österreichs, VTÖ): It co-ordinates a network of about 30 innovation, entrepreneur and technology transfer centres, widely acknowledged as being an important instrument of regional policy. The most prominent centre focussing on environmental issues is the 'Cleaner Production Centre Austria' in Graz. It is supported by the city of Graz and the ÖKK (see below).

R&D activities in Austria are predominantly carried out by university institutes proper or university-associated research units as well as large private (usually non-profit) specialised research centres. The two most important institutions among the latter are located near Vienna (Seibersdorf research and Arsenal research) and in Graz (Joanneum Research).

- *Austrian Network of Environmental Research* (NUF): This network, co-ordinated and partly financially supported by the Federal Ministry of Science and Transport, was created to put environment-related research in Austria into an international context. It aims at the promotion of interdisciplinary research in this field among national research institutions as well as the participation in international projects.
- *The Österreichische Kommunalkredit*, ÖKK, (Austrian bank for loans to local administrations) founded as a specialised bank in 1958 is authorised by the 'Environmental Support Act' (1993) to handle Austria's environment-related financial support activities. The original purpose of this institution was to support long-term investment projects, mainly in material infrastructure, by local authorities. The long involvement in water management projects, a typical investment by communities led to a long learning process concerning environmental issues. This experience made the ÖKK an ideal agent for all environmental policy activities involving investments and government subsidies in the nineties. This institution is now charged with the following environment-related tasks:
  - Environmental support in Austria and abroad.
  - Support of water management activities.
  - Rehabilitation of contaminated sites.

Due to the importance, closeness to the addressees of policy and its well developed service character, as well as the large volume of activities of public sector programmes the ÖKK is handling, it exerts a considerable influence on environment related technology policy in Austria.

The following table provides a short overview of the federal Austrian institutions funding technological innovation activities.

Table 2.1: Federal Austrian environment-oriented technology policy funding institutions

Funding Institution	Supervising Ministry	Year of establishment	Relevant act	Tasks
ÖKK	BMU	1958	Environmental Support Act 1993	Environmental support
FWF	BMWV	1967	Research Funding Act 1967	Basic research funding
ERP	BMWV	1962	European Recovery Programme (Marshall Plan)	Direct research funding, interest supported loans
ITF	BMWV/ BMwA	1987	ITF Act 1987	Support of national and international research and development projects, interest supported loans (managed by ERP and FFF)
FFF	BMwA	1967	Research Funding Act 1967	Individual project financing
BÜRGES	BMwA	1954		Funding of SME

2.3.2.3   Programmes and policy areas
European Union initiatives
– PREPARE (Preventative Environmental Protection Approach in Europe, since 1991): It is an initiative in the framework of the EUREKA programme to support enterprises to develop measures to avoid waste and emissions. The overall objective of PREPARE is to identify the need for, and stimulate the initiation of new international, industry-oriented R&D projects in the field of cleaner process technologies and cleaner product development. In Austria the Federal Ministry of Science and Transport and the Federal Ministry of Environment were responsible for the realisation of the projects. Meanwhile the initiative was extended to the regional level, in Styria, Upper and Lower Austria, Carinthia and Vorarlberg PREPARE projects are under way. The PREPARE approach is based on the combination of management and process engineering procedures.
– LIFE (since 1992): It is a financial instrument for three major areas of action, namely Environment, Nature and non-EU countries. LIFE is implemented in several phases, the second phase is running from 1996 to 1999 with a total budget of €450 million, 46% of which is earmarked for LIFE-Environment projects. 'LIFE-Environment' finances preparatory, demonstration, technical assistance, support or promotional measures designed to:
  – Promote sustainable development and integration of the environment in industrial activities;
  – Help local authorities to integrate environmental considerations in land use development and planning;
  – Strengthen the link and complementarity between environmental regulations and structural financial assistance, in particular from Community funds and financial instruments concerning the environment.

Federal initiatives
– 'Environmental Support in Austria' (since 1993, 'Austrian bank for loans to local administrations', ÖKK): It provides financial support (approx €28 million per year) for enterprises realising environment related measures. Since 1997, the Federal Ministry of Environment has placed a special emphasis on the prevention of climate change, with 'energy conservation/energy efficiency' and 'substitution of sources of energy' constituting the special priorities of its support policies.
– Programme to Promote Sustainable Economic Activities (since 1999, Federal Ministry of Science and Transport (see 2.2.2)

Regional initiatives
- EcoProfit (Ecological Project for Integrated Environment Oriented Technology, since 1991): The first EcoProfit project was carried out in the city of Graz. Meanwhile more than 150 enterprises in 11 cities (regions) have participated. The core of the project is an intensive co-operation between companies, administration and research units with the intention to initiate the introduction of voluntary environmental measures, among these the implementation of measures involving cleaner technologies. The projects aim particularly at SMEs providing incentives (financial support for the major part of the project management, the promotion work and the consultancy by technical experts) for voluntary action to improve environmental performance. The main advantages of the participation in an EcoProfit project for the companies are:
  - Making use of the advice by technical experts at low costs and profiting from the experts´ knowledge of the state-of-the-art as a consequence of the consultants´ professional experience in universities and research institutes;
  - An improvement of relations with the local administration;
  - Utilisation of ecological and economic potential to save resources;
  - Widening and diffusion of environment-related innovation (Huchler and Martinuzzi, 1997).
- Regional Clusters: The Austrian Technology Policy Concept (draft of 1994) identified 'clusters' (i.e. networks of business firms, education and research units that work together on the development, manufacture and maintenance of products) as an important tool of a diffusion-oriented technology policy. The aims of the formation of clusters are the strengthening of the regional economic potential and competitiveness and the opening of new markets. Due to the deficit of R&D activities and the limited technology intensity of the economy this concept is developing relatively slowly. Meanwhile several clusters (Automobile-, Ceramics-, Wooden Furniture, Textile Industry) can be identified. Recently several environment-oriented clusters have been developing, e.g. the Austrian biomass cluster, the aim of which is the pooling of the existing competences in the field of biomass utilisation and the creation of a brand name 'Bioenergy Austria'. The entire region of Styria is attempting to create an 'eco-cluster', i.e. a network of companies dealing particularly with alternative sources of energy, new environmentally compatible production inputs as well as 'bio-architecture'. The network is supported to promote technological, management and marketing co-operation.

2.3.3 Evaluation of environment-oriented technology policy in Austria

Environment-oriented technology policy as a conscious effort to promote technology development and diffusion has been in existence for about three years. It is too early to pass final judgement on the success and shortcomings of this new area of policy. The following statements, mostly the results of expert interviews are hence only of a very tentative nature. Very few studies exist (e.g. Huchler and Martinuzzi, 1997) that shed some light on this policy area, the first evaluation studies are now under way.

A few general remarks about the state of TP in Austria are necessary to provide the required context and backcloth.

Since the early 1990s Austria's ratio of research and development expenditure to GDP is stagnant at a level of about 1.5% (OECD average 2.15%, EU average 1.85%). The public expenditure on research and development is close to the OECD average, private enterprise expenditure is clearly below (Hutschenreiter et al., 1998).

Referring to patent intensity (number of per-capita patent applications) Austria ranks in the lower middle group of the European countries. Furthermore, Austrian patent applications are highly concentrated in the context of the 'construction' sector, high-technology areas such as electronics or communications play a minor role (Hutschenreiter et al., 1998). Contrary to many other countries the military sector in Austria is only of negligible importance.

In comparison to the overall deficits just mentioned, money spent on environment-related R&D and diffusion has constituted a significant share of total expenditures. The ITF fund for example spent 14% of its budget on environment related technologies although the absolute magnitude of € 3.49 million is not overwhelming. This amount also lies considerably below the sum spent for environment related research projects laid out by the 'Fund for the Promotion of Scientific Research' (€13.66 million in 1995), which relies on applications from the research community (initiatives from 'below'). Similarly the 'Austrian bank for loans to local administrations' (ÖKK) spent €37.5 million in 1995 to support the installation of new environment related technologies.

The expert interviews conducted yield a few tentative first insights into the strengths and weaknesses of the new environment-oriented technology policy programmes.

Strengths:
– The researchers active in this field are highly qualified and competent.

- The co-ordination between federal funding institutions (especially since the 2nd Technology Policy Concept, 1996) has improved considerably.
- The recent concentration of funding into focussed targeted research programmes is likely to lead to greater success than the rather dispersed funding practice hitherto.
- The complementary funding opportunities by EU programmes (since 1996) have greatly increased the chances as well as international competitiveness of the Austrian R&D community.

Weaknesses:
- A future lack of competent R&D personnel is envisaged. The present situation, particularly in the private R&D sector, which is excellent, is partly due to the temporary immigration of East Europeans. With the accession of the home countries to the EU this could change dramatically.
- Lack of experience in formulating project proposals (especially with regard to EU-projects).
- Referees of 'Science Foundations' tend to be rather conservative (difficulties for interdisciplinary and high-risk research proposals to be accepted).
- Major controversial new research fields remain uncertain candidates for Austrian research funding, as pertinent political decisions are not made (e.g. genetic engineering).
- R&D and implementation programmes still lack co-ordination, hence know-how transfers are still 'hesitant'.

2.4 Interlinkages between environmental policy and environment-oriented technology policy: Strengths and Weaknesses

There is a consensus among experts that traditional environment related technology (mostly end-of-pipe) has been successfully installed in Austria at a broad scale. This success is partly due to the stringent command and control system prevalent in environmental policy and a substantial volume of voluntary investment by companies based on a fairly widespread 'green conscience' among entrepreneurs and households as well as voluntary covenants in some sectors of the economy.

It is frequently argued, however, that there has been a lack of explicit co-ordination between environmental policy, research and technology adoption programmes until the early 1990s. In the 'alternative sources of energy' field, e.g. bio mass use research was supported but incentives for diffusion of the technologies developed were not provided sufficiently by environmental policy. Furthermore the lack of special earmarked programmes for environment-oriented R&D and particularly technology diffusion policies were lacking.

In the mid-1990s the Austrian economy was hit by the worldwide recession leading to increasing competition on global markets. When Austria joined the EU in 1995 the pressures on the country's economy grew, so did the chances to open up new markets. It was recognised, however, that the products made in Austria were on average not as knowledge-intensive as those of the countries in the same GNP/capita league and that considerable efforts to promote R&D in Austria were required. Additionally the situation was aggravated by the fact that the Maastricht treaty demanded a reduction of governmental budget deficits thus making large public expenditure prone programmes difficult to establish.

An attempt was made to increase the efficiency of all the policy programmes, specifically in the environment-related R&D field. One strategy was to create integrated foci for research in response to policy priorities. As a consequence an environmental action plan was approved by Parliament in 1995 (National Plan for the Environment, NUP), and consistent with it, a new technology policy concept was adopted. Classical TP and the new Environmental Policy Concept thus led to a new explicit policy field, i.e. environment-oriented technology policy. With respect to environmental goals special earmarked programmes were created (see 3.2), aiming primarily at the creation and diffusion of 'cleaner technologies', which were to be beneficial to the environment as well as to product quality and cost effectiveness and thus also for Austria's economy. The following overview shows the principle linkages between the relevant policy fields and socio-economic stimuli and the responses by companies since the start of the programmes.

Figure 2.4 shows the stimuli active in Austria's environment-oriented innovation activities since the mid 1990s.

As becomes evident from the short description of the relevant programmes in section 2.3.2, there is still a multitude of initiatives which are not integrated into a single streamlined environment-oriented technology policy concept. Depending mainly on the organisations standing behind the various initiatives the foci of these vary. There is, however, some overlap.

The early stages of a full innovation cycle, i.e. R&D and invention are still supported by the institutions founded in the late 1960s (much as the equivalent of a National Science Foundation, etc., see 2.3.1). New financing opportunities for Austrian researchers have existed since the recent accession to the EU, opening the various environmental research programmes to applications from interested scientists.

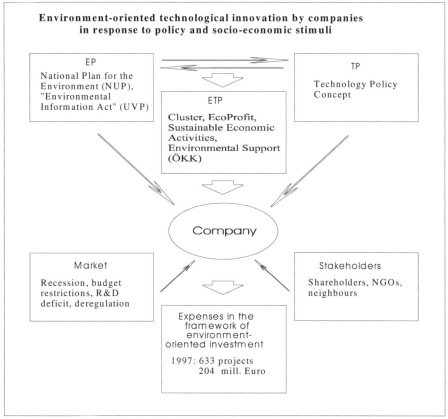

Figure 2.4 Environment-oriented technological innovation by companies in response to policy and socio-economic stimuli

The more recent programmes emphasise the integration of R&D with the technology diffusion phase. The initiative on 'Sustainable Economic Activities' was developed and is backed up by the Federal Ministry of Science and Transportation and has just been launched (1999) issuing the first calls for innovative projects. The core of the programme is certainly the link between environmental goals stated in the National Plan for the Environment (NUP) and an economically viable technology development and diffusion policy, explicitly introducing market criteria into the process aimed at.

Due to the experiences made in the past, as already mentioned, the adoption and diffusion of environmental technology has gained importance. This trend is clearly demonstrated in the programmes EcoProfit, PREPARE as well as the new Energy Concepts, where R&D plays no explicit role and the emphasis is on an operationalising knowledge transfer (mostly via consultancy) and technical and organisational change in companies.

Regional initiatives, by the very nature of a federal country such as Austria, have always played a major role with respect to environmental issues (green initiatives) as well as local economic development. The recent Technology Policy Concept aims at the creation of 'Regional Clusters'. In some cases explicit 'Environmental Technology Clusters' are targeted (e.g. in Styria), but environmental objectives are always on the list of support criteria for all clustering initiatives, thus creating potential indirect positive environmental impacts.

Table 2.2: Recent environment related programmes and policy areas

Focus of the initiatives	Programmes & Policy Areas	Environmental Impact	Goals	Strategies		Instruments	Strength of Stimulus (ranking)			
				Management	Technical		1	2	3	4
Economic	Regional Clusters	indirect	strengthening regional economic potential competitive-ness opening of new markets	creation of regional innovation system	end-of-pipe, recycling, measure technologies	financial support for: business start ups, consultancy incubation & innovation centres	M	TP	S	EP
Diffusion	Sustainable Economic Activities EcoProfit PREPARE Energy concepts	direct	diffusion of environmental technologies compliance with international standards	co-operation networks know-how transfer	clean technologies, renewable resources	environmental consultancy, environmental awards	EP	TP	S	M
Innovation	Sustainable Economic Activities Env.Support LIFE	direct	clean technologies BAT 'best practice'	single projects	clean technologies prototype	tenders single project subsidies	EP	TP	M	S

M = Market Forces; TP = Technology Policy; EP = Environmental Policy; S = Stakeholders

The programmes described are also characterised by some novel approaches to the management methods. Considerably more emphasis is placed upon co-operation between the relevant actors (including environmental control authorities, technological and management consultants as well as the financial support institutions, private and public). It is also worth noting that the scope of instruments has become wider and is not limited to subsidies any longer, but includes knowledge transfer centres, network formation, awards and competitive tenders for financial support. These new environment-oriented technology policy programmes were created in response to stimuli provided by the market forces as well as traditional policy areas (such as regional policy, environmental policy and TP). The stimuli exerted by these forces of influence varied in intensity, but environmental and technological considerations are dominating this policy era. The following table 2.2 provides an overview, including a tentative ranking of the significance of various policy fields.

2.5   Summary and future challenges

The evolution of a system of environmental policies started in the 1970s with the creation of a ministry charged with the task to establish a legal framework as well as to provide a milieu to solve Austria's pressing pollution problems. Most of the direct responsibilities for the management of environmental resources remained in the administrative realm of the various sectorally organised federal ministries as well as the regional and local control authorities. A system of 'command-and-control' regulations was instituted in response to the most pressing needs and political pressures exerted by green grassroots movements and increasingly NGOs with environmental purpose. The heydays of the development of this policy system and the most spectacular events in this area occurred in the 1980s, since in the 1990s the waters became calmer and the dealings with environmental issues less emotional and more professional. This policy system can clearly look back at various success stories, making Austria a very 'green country' in the European context.

Despite the fact that the important role of technology to combat (originally mostly industrial) pollution has been recognised from the early days on (as demonstrated by the creation of a very important 'behind the scene' type of discussion and policy preparation group, the 'Austrian Association for the Environment and Technology' - ÖGUT), no coherent, targeted policy area was set up. Environmental technology development was just seen as a part of Austria's warranted effort to overcome a general R&D deficit. Despite this fact environment-oriented research activities took a significant share of the public funds dedicated to the objective to become more competitive. In the early 1990s an attempt was made to utilise this research boom to promote a presumed international market niche in environmental technology (predominantly end-of-pipe) made in Austria, which did support parts of the former nationalised industries in the transition to modern private industry at a large scale.

In the mid-1990s criticism of the prevalent environmental policy and TP policy systems became more prominent in the media and the public including some of the experts working in the policy execution institutions. It was realised that in an era of shrinking government budgets and a drive towards privatisation, the policy style had to change and new, more efficient instruments needed to be introduced and new programmes to be designed. The motto was now that the command and control system had to be complemented and partly replaced by incentives and approaches based on voluntary participation deemed to be less intensive in costs of public transactions and much better in line with market processes. At the same time it was realised that an integrated, targeted environment-oriented technology policy had to be developed, spanning the whole spectrum of the phases of an industrial innovation process. Particularly, it was requested that the diffusion of new technology be more emphasised than hitherto. In the second half of the 1990s such innovation-oriented programmes of action were introduced, taking on board many of the postulates developed in innovation studies, such as the integration of all relevant actors in the innovation process by facilitating or even requiring the establishment of networks, as a necessary condition for financial support. It is too early to judge whether these new programmes will lead to the desired environmental and economic effects as the time of implementation is too short, but some challenges for the future can be tentatively identified.

Despite the fact that the new approach is intended to overcome a situation of fuzziness and intransparency of responsibilities of different tiers of government and their various agencies, more efforts in this direction are still warranted. This task is certainly not facilitated by the fact that the EU level has now entered and continues to do so increasingly into Austria's three tier government system typical of a federal country. The trend towards deregulation and administrative efficiency, threatening the dominance of the command and control system, is going to be quite a chore for the legislative bodies in the country.

Additional steps in the direction of making sure that the important actors in environment-related innovation processes promoted by public money are on board seem to be necessary in the future. As the private sector is targeted to play a greater role in this policy area in the future, particularly the financial intermediaries need to be integrated into the evolving networks of relevant actors. The privatisation of formerly public management tasks will have to include the control of emissions and other environmental impact prone activities of industries as well as households. The international experience available in this field is not abundant, yet, proactive activities and research along these lines as well as gradual inclusion into existing policy programmes will be badly needed.

The environment-related R&D capacity has developed favourably over the last two decades, but the future of the entire research sector in Austria needs to be carefully considered and planned. The deficits in the Austrian university systems are now a heated political discussion issue and reform efforts are under way. An often neglected aspect is the evolution of a European system of higher education and academic training, which will require the positioning of the R&D institutions in a European (and increasingly global) system. The environmental sector could certainly, given its present state of competence, networking experience and development potential be targeted as one of the focal areas of support and promotion.

Environmental policy is facing two major challenges in the future. The climate-related international agreements, particularly the Kyoto protocols, have introduced an old theoretical concept into political reality, i.e. the necessity to define quantitative contingents of maximum allowable emissions. In terms of the implementation of these concepts a new management perspective is absolutely necessary and not many international learning experiences can be relied on. Approaches such as the emission trading system in the U.S. most likely require new thoughts on compatible and integrated environment-oriented technology policy programmes.

REFERENCES

Aiginger, K. (1997) Industriepolitik. In:H. Dachs et al. (eds.) *Handbuch des politischen Systems Österreichs*. Wien: Manz, pp. 557-566.

BMWF (1994) *Technologiepolitisches Konzept 1994 der Bundesregierung. Expertenentwurf.* Wien.

BMwA (1998) *Leistungsbericht 97/98. Umweltschutz und Umwelttechnik.* Wien

BMWF (1989) *Technologiepolitisches Konzept der Bundesregierung und Katalog operationeller technologiepolitischer Maßnahmen.* Wien.

BMWV (1996) *Technologiepolitisches Konzept der österreichischen Bundesregierung 1996.* Wien.

Bundesgesetz zur Förderung der wissenschaftlichen Forschung (Forschungsförderungsgesetz, FFG), BGBl. 377/1967.

Galley, H. (1997) Regionalwirtschaftliche Impulse durch Technologie-, Innovations- und Gründerzentren. In: *Wirtschaftspolitische Blätter* 5/1997. Wien.

Glatz, H. (1995) *Österreichische Umweltpolitik. Eine kritische Einschätzung der Instrumente.* Wien: Kammer für Arbeiter und Angestellte für Wien.

Gottweis, H., Latzer, M. (1997) Technologiepolitik, in: H. Dachs et al. (eds.) *Handbuch des politischen Systems Österreichs*. Wien: Manz pp. 652-663.

Hofer, R., Hutschenreiter, G., Polt, W. (1998) Technologie- und Innovationspolitik als Industrie- und Beschäftigungspolitik. In: Zukunfts- und Kulturwerkstätte (ed.) *Re-engineering der österreichischen Industriepolitik*. Wien, pp. 20-51.

Huchler, E., Martinuzzi, A. (1997) *ÖKOPROFIT Dornbirn*, Schriftenreihe des Bundesministeriums für Umwelt, Wien.

Hutschenreiter, G., Knoll, N., Paier, M., Ohler, F. (1998) *Austrian Report on Technology 1997*, Wien.

Kanatschnig, D. (1986) *Präventive Umweltpolitik. Gestaltungsprinzipien der Vorsorgeplanung*. Linz: Universitätsverlag Rudolf Trauner.

Köppl, A., Pichl, C. (1995) *Wachstumsmarkt Umwelttechnologien. Österreichisches Angebotsprofil*, Wien.

Kuntze, U., Köppl, A., Pichl, C. (1997) *Wirkungen der Innovationsförderung im Schwerpunkt Umwelttechnik des Innovations- und Technologiefonds* (ITF), Karlsruhe und Wien.

Lauber, V. (1997a): Umweltpolitik. In: H. Dachs et al. (eds.) *Handbuch des politischen Systems Österreichs*. Wien: Manz, pp. 608-618.

Lauber, V. (1997b) Austria: a latecomer which became a pioneer. In: M.S. Andersen, D. Liefferink (eds.) *European environmental policy. The pioneers*. Manchester and New York: Manchester University Press, pp. 81-118.

Martinsen, R., Melchior, J. (1994) *Innovative Technologiepolitik. Optionen sozialverträglicher Technikgestaltung – mit einer Fallstudie über Österreich*. Pfaffenweiler: Centaurus-Verlagsgesellschaft.

Österreichische Bundesregierung (eds.) (1995) *Nationaler Umweltplan* (NUP). Wien.

Rothschild, K. (1989) Ziele, Ereignisse und Reaktionen: Reflexionen über die österreichische Wirtschaftspolitik. In: H. Abele, E. Nowotny, S. Schleicher, G. Winkler (eds.) *Handbuch der österreichischen Wirtschaftspolitik*, 3. Auflage, Wien: Manz. pp. 113-123.

Schramm; W. (1994) *Technologiepolitische Schlußfolgerungen des Technologie- und Umweltprogrammes der OECD für Österreich*. Wien: Österreichsiches Akademie der Wissenschaften.

Schremmer, C., Tödtling, F. (1996) *Regionale Industriepolitik für Österreich*. Wien: Österreichisches Institut für Raumplanung.

Streissler, E., Neudeck, W. (1997) Wachstums- und Umweltpolitik. In: E. Nowotny, G. Winckler (eds.) *Grundzüge der Wirtschaftspolitik Österreichs*. 2. Auflage, Wien: Manz, pp. 166-213.

Tajmar, P. (1998) Sauber produzieren. In: *industrie*, Nr. 3. Wien.

Traxler, J. (1993) *Technologieorientierte Industrie- und Gewerbepolitik*. Working paper, Wien.

APPENDIX

Existing sectoral agreements:
- Used tires: about 50,000 tons in 1997
- Car batteries: about 18,000 tons in 1997
- Other batteries: between 2,100 and 2,600 tons in 1997
- PVC windowframes: 50 tons
- Plastic pipes: 381 tons collected in 1997
- PVC-floor coverings: about 100 tons collected in 1997
- PVC credit, ATM identity and club membership cards: 4.62 tons collected in 1997
- Blister packing materials for drugs: 101.7 tons collected in 1997
- Used glass: 239,439 tons collected in 1997
- Used paper: about 1.1 million tons in 1997
- Beverage card-board packings: 14,856 tons collected in 1997
- Used textiles: about 20,000 tons per year
- Electronic waste materials: between 80,000 and 90,000 tons in 1997
- Recycled used cars: 91,000 in 1997

- Construction refuse: about 20 million tons in 1997
- Imports of tropical wood: tropical wood is only imported if originating from an area with a sustainable forest management system
  (source: BMwA, 1998)

Interview partners:
- BERGER-HENOCH Berthold, Federation of Austrian Industry
- HACKL Wolfgang, Austrian bank for loans to local administrations (ÖKK)
- LAUBER Wolfgang, Chamber of Labour
- OSTERAUER Michael, Federal Ministry of Foreign Affairs
- PAULA Michael, Federal Ministry of Science Transport
- SCHNITZER Klaus, Austrian Industrial Research Promotion Fund
- SCHWARZER Stephan, Chamber of Commerce

Chapter 3

Environmental Policy and Environment-oriented Technology Policy in Denmark

OLE ERIK HANSEN, JESPER HOLM, BENT SØNDERGAARD
Department of Environment, Technology and Social Studies, Roskilde University, Denmark

3.1   Introduction

In Denmark the executive power of the government has been administered under a ministerial system since the first constitutional act of 1849. Today, the Danish government encompasses 18 ministries. The two most central ministries with respect to environmental policy (EP) and technology policy (TP) are the Ministry of Environment, established in 1971, and the Ministry of Trade and Industry, which has existed since 1908. A general industry policy to promote certain structural and technological inventions and innovations has existed in Denmark to various degrees since the beginning of the twentieth century. Several R&D, information, and standardisation institutions were established during the last century along with a number of taxation, subsidy, and service schemes (Christiansen, 1988). However an *explicit* policy area concerning TP did not come into being until the beginning of the 1970s, and an *environment-oriented TP* was first launched in 1985-1986. In terms of an EP, some health orders and regulations over nuisance and discharges into urban watercourses date back to the middle of the nineteenth century, the first law on nature preservation was passed in 1917, and the first act concerning urban planning and development was passed in 1925. However, a genuine policy on environmental protection started as late as 1972 with the first Environmental Protection Act, and the establishment of the Ministry of Environment that covered all fields of EP. Since then, a comprehensive set of regulatory, monitoring, research, and

59

Geerten J.I Schrama and Sabine Sedlacek (eds.) Environmental and Technology Policy in Europe.
Technological innovation and policy integration, 59-96. © 2003 Kluwer Academic Publishers. Printed in
the Netherlands.

subsidising institutions have been developed within the ministry. A consensus-oriented policy style ensured an economic-technological concern in environmental regulation from the start, as Danish industry and trade organisations were consulted on standards and the enactment of new initiatives. Nevertheless, it could be argued as to when a *deliberate* technology orientation for EP started. Besides an investment scheme for purification treatment utilities in 1975, it was only in 1985-86 that a number of R&D and innovation programmes on clean technology were started. Accordingly, we will, in what follows, focus on EP and TP and their respective crossover since the mid 1980s.[1]

3.1.1   Central-local authority patterns

There are three levels of public administrative competence with respect to Danish EP and TP: the ministries, the counties, and the municipalities. For TP, the municipalities and the counties play a minor role, but they co-operate with very important regional semi-public R&D institutions for technology innovation and networking. In terms of EP, Denmark is characterised by a significant delegation of implementation and administration of the Environmental Protection Act to local authorities. Therefore a brief outline of the local administrative structure is provided.

It is the national Parliament (Folketinget), the Government and its ministers, that regulate the majority of tasks and the primary revenue base of the local authorities, according to sector-specific acts (environment, transport, housing, taxation, and finance). Within this framework, local authorities are free to take their own initiatives and find their own financial resources, although they are restricted by a general, and legally not very clearly defined, set of authoritative codes of conduct (*kommunalfuldmagten*). Even though local authorities carry out some 70 per cent of national state business, their primary revenue stems from local income tax and property taxation. The state provides direct subsidies linked to certain state initiated tasks and regulate an inter-municipal economic transfer scheme according to

---

[1] When we discuss technology policy (TP) and environmental policy (EP), and the crossover between the two, we may distinguish between *deliberate* efforts to foster certain technological innovations for the benefit of the environment, from innovative technological *side effects* from the EP, or environmental beneficial side effects from TP. We will primarily focus on the first category, but will give some descriptions of the technological impacts of the various regulatory strategies within the Danish EP. Secondly, it is problematic narrowing the focus down to these two ministries, as a significant increase in interpolicy or cross-sector policy efforts have taken place since beginning of the 1980s. Thus EP has diffused into the policy sectors of housing, food, education, and taxation, energy, transport, and agriculture. These efforts will be described briefly, but the analysis will not go into the general policies of these ministries.

differences in economic and social indicators. Since the beginning of the 1970s, *the counties* have, in relation to the environmental sector, been responsible for:
- nature conservation and restoration;
- control of the environmental quality of drinking water, streams and fjords;
- regional master (spatial) planning; and finally
- environmental control and servicing of larger enterprises.

Related to the environmental sector, *the municipalities* have, for their part, responsibility for:
- control and licensing of small and medium sized enterprises (SMEs);
- municipal spatial zoning;
- public water works;
- the gathering and handling of waste;
- waste water planning for all streams;
- public incineration utilities and waste deposit sites;
- sewers;
- energy transmission; and
- sludge treatment.

In 1969-70 a major municipality and budget reform process laid the basis for the current political administrative structure, consisting of 14 counties and 275 municipalities, which vary considerably with regard to the number of inhabitants and socioeconomic structure. The reforms reduced the number of authorities considerably, in order to strengthen the political, administrative and economic role of the municipalities and counties. Thus a centralisation of the local administrative apparatus has conditioned a decentralisation of central state control during a period of general growth in the public sector. Ever since the economic crisis of the mid 1970s, that gave new strength to local economic policy interests, centralisation and decentralisation have become a central topic on the political agenda. The debates have concerned burden sharing and the separation of political and economic responsibility. Even though neo-liberal political ideologies have dominated the agenda since the beginning of the 1980s, it has been vital for all Danish governments to maintain a welfare-based political legitimacy. The dominant policy style in Parliament and in local government has been flexible; the sector ministers and the government govern by guidelines and negotiations, leaving space for various local interpretations.

For the EP sector, this development has resulted in a continuous political-administrative conflict over the burden sharing of environmental costs and duties between central and local authorities. Whereas, initially, the environmental regulation of industry and agriculture were decentralised, the last twenty years have seen the development of various forms of centralisation. In the TP sector, the strengthening of the local authorities in the 1970s paralleled the establishment of a more firm and specific TP policy discourse at the national level. Due to the local economic problems of rising unemployment, social costs, and loss of tax revenues since the mid 1970s, the municipalities have started a number of technology and business sector initiatives. These initiatives had a dual perspective: to mobilise along innovative development paths the companies already located, and to site new companies by drawing attention to favourable manufacturing conditions.

3.2   Environmental policy in Denmark

3.2.1   Institutional characteristics

The establishment and institutional division of a comprehensive Danish political *sector for environmental regulation* took place in 1971-76. Even though there have been many changes in office and departmental structures, the basic principles have remained. Today, the Ministry of Environment and Energy (MEE) is organised with a central department serving the minister and the government. Three agencies[2] take care of:
– identifying polluting activities; monitoring nature and environmental conditions;
– consulting and networking with target groups;
– making new inventories;
– drawing up strategic planning documents; and
– preparing R&D and law initiatives.

Further, the agencies have established three strategic research institutions[3], covering environmental monitoring and indicator development, assessments of impacts from public inventions, and forwarding proposals for political priorities to the minister. A planning department covers all aspects of spatial planning, including preparation of the national Mandatory Planning Documents. Today, the number of administrative staff at the Ministry is approximately 1500 (Moe, 1997).

---

[2]   Danish Agency for Environmental Protection (DEPA), the Danish Energy Agency, and the National Forest & Nature Agency (resources, forests, cultural heritage, natural areas).
[3]   The National Environmental Research Institute, the Geological Surveys of Denmark and Greenland, and the Danish Forest and Landscape Research Institute.

The central institution within the ministry related to industrial environmental protection, and thus technological innovation within business, is the *Danish Environmental Protection Agency* (DEPA*)*. The focus and structure of the expert offices within this agency have changed over time, and some cross-sector offices have been established (such as the Offices of Economy, Staff, Communication, and Control and Law). However, the division of labour among the expert advisory offices reflects a functional differentiation relating to:

- pollution media (wastewater, open sea, soil, and water intake);
- problematic substances (pesticides, chemicals, waste);
- target areas (industry and transport);
- add-on activities following the historical development of new interventions (the Offices for Cleaner Technology, for Environmental Aid, and for Environmental Exports).

This institutional arrangement of offices has created various hindrances for the paradigmatic renewal of politics. The central areas of industry and technology relevant initiatives within the DEPA are:

- Identifying pollution or hazardous activities, producing guidelines and regulations for these activities, and controlling the application of the guidelines by local authorities. In terms of research and technology development, the agency is heavily subsidising R&D units among consultants, universities, and business R&D units.[4]
- Supporting research and diffusion of cleaner technologies and products. Currently the fourth subsidy programme since 1987 is supporting R&D, diffusion, and competence building for firms and R&D units involved in cleaner product/technology development.
- DEPA gives consultations and negotiates with a vast number of trade organisations, local authority associations, and green non-governmental organisations (NGOs) on strategies, agreements, and enactment.

According to a political-strategic organisational conceptualisation, the *institutional* development of the ministry has gone through three stages (N.A. Andersen, 1995: 181-197). These stages coincided roughly with major phases in the overall strategic attempts of the Danish environmental policy towards industry (Christiansen and Lundquist, 1996; Andersen and Liefferink, 1997; Schroll, 1997). Let us therefore give a short overview:

1. 1972-1984: *Institutional set-up:* attempts to establish a priority and top-down strategic planning model for integrating sector planning into an overall planning division under the department (the Office for Law

---

[4] The latter covering for example Akademiet for tekniske Videnskaber, Teknologisk Institut, Institut for Produktudvikling.

Preparation and Co-ordination). In other words, a profoundly political and administrative decentralised structure of implementation – based on municipalities and counties. *Regulatory strategy*: the establishment of guidelines and practices for a hierarchical set of ambient environmental quality monitoring and standards. This formed the background for optional non-binding *guidelines* for effluent standards, finally used for issuing discharge permissions to firms. Subsidies for those who lacked the capital for investments in environmental protection.

2. 1984-1990: *Institutional set-up:* a modernisation process under liberal-conservative governance, with strategic efforts to frame priorities under constrained economic conditions; forwarding ministerial capacity for alteration by strategic heading for a responsive culture, and delegating responsibility for new issues and negotiations to the agencies. In addition, a partial centralisation of implementation duties to counties started, along with a rationalisation of the permission system by differentiating among the target groups. *Regulatory:* building infrastructure for pollutants, forwarding mandatory effluent standards, ideologically aiming for extra-legal efficiency. Initiating comprehensive environmental action plans for other policy sectors.

3. 1990 onwards: *Institutional set-up:* the ministry has tried to position itself as a new cross-sector interpolicy ministry, by fostering an eco-societal discourse on the government's joint effort to co-ordinate ministerial policies. The ministry has deliberately strived to influence the public agenda, the other ministries, and the government by forwarding new eco-socio-oriented images of society, and initiating strategic environmental impact assessment (SEIA) procedures for all proposals and planning in other relevant ministries. It has also been fostering new interactive, flexible, relationships between environmental authorities and business. *Regulatory:* the environmental policy has integrated Life Cycle Assessment (LCA) and Best Available Technology (BAT) orientation into legislation, forwarded product stewardship, and launched new efforts for diffusing cleaner technologies. New voluntary, market-based and interactive instruments have been introduced accounting for business responsibilities in environmental policy (voluntary agreements, EMAS, Green Accounting, public green purchasing). Finally, strategic planning was introduced with the successive use of environmental indicators, SEIA of plans, and with the introduction of a quarterly by giving priority to special issues.

As described above, the local administrative institutions involved in environmental regulation are the 14 counties and the 275 municipalities. In relation to industrial pollution prevention they are obliged to enforce the law, establish infrastructure, and monitor environmental quality. They are also free to take other initiatives in the environmental field. The controlling and permitting staff for industry and agriculture within the local environmental departments number 300 for the counties, and 450 for the municipalities (Miljøministeriet, 1995). The Danish consensus-seeking and neo-corporate environmental policy style has fostered a governing inclusion of a large number of Danish organisations for industry, agriculture and trade; organisations for energy suppliers; waste handlers; national associations for local authorities; and a few green NGOs.

3.2.2   The standard environmental regulation towards industry

3.2.2.1   Legal setting
The fundamental laws covering industry-related environmental protection are the Environmental Protection Act (1999), the Spatial Planning Act (1998), the Act on Nature Protection (1999), the Act on Chemical Substances and Products (1999), and the act on GMO (1992). The rules and regulations stemming from the Environmental Protection Act (first enacted in 1973) will be discussed below, as this is the main regulatory framework for environmental behaviour of industries. The act on Chemical Substances and Products (enacted in 1979) is also of major importance for the environment, but it has not yet had a major innovative impact on industry. This act established a centralised administrative system, requiring mandatory approval of new toxic substances and products, labelling of existing chemical products and restrictions on the use in industrial processes and products according to health and environment effects. The law has developed through a number of orders, mainly due to EU legislation, but also with some restrictions related to the precautionary principle (bans on cadmium, lead bullets, CFCs in coolers, Pentachlorophenol, creosote, and several pesticides).[5] Recently, the Minister of Environment launched an early warning about 100 unwanted toxic substances that the ministry would like Danish industry to avoid. To a certain extent, the Planning and Nature Preservation Acts form the background for the targets and standards set in industrial environmental regulation.

The first approach for regulating industry's environmental performance stemmed from *spatial environmental planning*. Since 1925, urban planning has had an important role in siting dangerous industry outside housing areas, but it was not until the comprehensive planning reform of 1969-73 that a firm and integrated planning system was developed for regulating industrial

---

[5] For Danish agriculture, the law has been used for a strict reassessment procedure for all the active substances in all pesticides on the market. The complicated assessment process and prohibitions have led to reducing the number of allowed active substances from 216 to 119 (Moe, 1997: 11-12).

siting in relation to a number of concerns. The planning reform encompassed three acts, (the *Act on National and Regional Planning* (1973), the *Act on Municipal planning* (1975), and the *Act on Urban- & Land zoning* (1969), establishing an all encompassing planning procedure with public hearings, which ensured a balance between national, regional, and local priorities. The spatial planning regulations reflected a dominant conception of environmental problems. The central objectives of this spatial planning reform were to separate industrial, agricultural, housing, and nature areas by zoning measures; to secure sustainable extraction of raw materials by zoning and licensing; to foster structured growth in urban areas; and to preserve vulnerable nature and water resources. Especially nuisance problems could be dealt with by a proper diffused use of nature's carrying capacity.

The economic crisis and neo-liberal deregulation politics, which began at the end of the 1970s to the late 1980s, ended urban, industrial and infrastructural expansion, and thereby removed most of the teeth of local spatial planning instruments. Furthermore, the formalised and progressive expertise-driven spatial planning system was reorganised in the late 1980s in order to give greater room for political manoeuvre to the minister and the government. The discussions concerning the renewal of the planning system and instruments continued during the 1980s deregulation period. This had a strong focus on the lack of legitimacy of public planning, and the need for greater involvement of citizens. Besides regulating landscapes, rivers, and nuisance; the spatial planning tools have become important in protecting important sources of drinking water from industrial and agricultural effluents, and have been used for fighting eutrophication of wetlands.

The Environmental Protection Act of 1973 was a framework-oriented and decentralised mode of regulation. This was both in accordance with the 1970s municipality and planning reforms that gave new political authority to municipalities and counties, but also reflected an understanding of pollution as a local nuisance, related to specific contexts. Accordingly, the local authorities, which were close to specific problems, were seen as to be best suited for control and conflict solving. The 1973 Act was a piece of comprehensive legislation that was intended to cover all types of pollution stemming from industry (agriculture was to come later).

3.2.2.2   *Instruments for environmental protection*
The measures favoured to regulate industry were the issuing of legally binding licenses, approvals and permits, and subsequent control by local inspectors. Local authorities were obliged to monitor environmental quality; run waste water treatment plants; resolve conflicts; enforce directives on waste handling, and thresholds for effluents. Implementation of legislation has taken place in two modes: in close co-operation with industrial and

agricultural associations, the Ministry of the Environment has had the role of filling in the framework by producing general guidelines, thresholds for controls, and servicing and the issuing of permits among local authorities. Secondly, the local authorities have been obliged to issue individual pollution and wastewater licences, control industry and agriculture, ensure nature preservation, draw up wastewater plans, and build solid waste sites and wastewater utilities. The regulatory approach to industrial pollution problems in Denmark was, and still is, in principal formed by the 'polluter-pays principle'. However, in reality, the public spending on incineration, waste water treatment, and solid waste handling, has covered quite a large amount of the costs.

3.2.2.3 Standards

Predominantly, regulation by law has occurred through the establishment of various standards, in order to demand a certain performance from industry. The regulation of industrial pollution problems in Denmark was initially based on an *ambient environmental quality* approach and case-by-case approval. Approvals, permits, or licences to heavily polluting industries were based on compliance with targets for *local ambient quality and neighbour nuisances*. Optional guidelines and thresholds for the ambient quality approach were, in principal, based on mapping and deciding specific environmental aims for the ambient quality of a specific recipient (city, stream, lake), and therefore differed from recipient to recipient. Thus, for water, the Danish Environmental Protection Act initially contained a demand for all counties to design *recipient quality plans* for lakes, inlets, and sea fjords in their regional spatial plans. These plans then had to be respected by the municipalities when surveying watercourses, drawing up wastewater plans, and running treatment installations. Thus, the municipalities were, theoretically, able to set standards for each individual company.

Initially with, the recipient strategy was related to industrial behaviour through the ambient environmental quality-oriented (AEQ) *standards*, that is concentration of pollutants in air, soil, and water, calculated and monitored based on specific dilution capacities of nature. This strategy focussed on *local* environmental problems and legitimised pollution to a certain extent by the use of judicial permits, or alternatively imposing dilution techniques (longer pipelines and taller chimneys) to secure corporate expansion. The ambient environmental quality strategy was also typically related to the use of various physical planning measures to concentrate industry in zones in which a great deal of pollution could take place (see further on). Whereas the protection of environmental quality is a central objective, this regulation also is built upon a legal principle of protecting industry from costly liability demands.

3.2.2.4   Outcome

The difficulties in calculating and setting effluent standards in relation to optional ambient environmental qualities, where tremendous and the recipient quality planning developed very incrementally. Accordingly, the counties' recipient quality planning system was suspended in 1986. The economic and political realities of the long economic decline, from the mid 1970s to mid 1980s did not encourage costly monitoring, and local authorities were more interested in attracting industry than in imposing limits on it. Accordingly, only a few counties fulfilled the act to the letter, and too often they made a very superficial and crude categorisation of recipients and so firms very easily obtained discharge permits (M.S. Andersen, 1995: 87). The socioeconomic result of this initial, but still prevalent, regulation has, in many cases, been resource-economic degradation at the expense of protecting the commons. The rather consensual and consultation-oriented policy style of the Danish EP, made sure from the start that corporate interests were present in new environmental protection efforts, so that standards were in line with technical and economic options. The *technological spin-offs* from the ambient quality strategy have been diverse dilution techniques, whereas no requirements were placed on the internal processes and products within the industry. For industry, the institutional set-up has typically resulted in rejection strategies or simple pay-the-bill attitudes. In the long run they have been misled into expanding until they hit the ceiling of the AEQ quota.

The optional regulation of industry, only with reference to ambient environmental quality standards resulted in an accumulation of ecological pressures. This has led to supplementary regulation since the late 1970s. These regulations have focused on the *pollutant infrastructure* and *end-of-pipe solutions* (filters, wastewater treatment units, incineration, and separation of hazardous wastes), reflected in new *emission standards* focussing on effluents from chimneys and pipelines. Some *guiding* emission standards for very toxic substances had already been introduced in 1974. However, parallel to a growing focus on *potential impacts* (rather than proven, calculated environmental quality or health impacts) emission standards became more common and also mandatory in the 1980s. Technically, the emission standards were related to toxic substances in effluents (in some cases also to toxic content in raw materials) and often calculated backwards from the capacity of various purification techniques. Local adoption of the standards through licenses and permits was secured by relating the general standards to a vast amount of specific manufacturing and monitoring conditions. For those industries that had difficulties in complying with the standards, a *subsidy scheme for environmental investments* was started in 1975 (Lov om støtte til miljøinvesteringer, M.S. Andersen 1995:

89) for installing purification treatment. This scheme was not meant to be innovative in terms of developing new technology, and was criticised for being used simply to gain profits[6]. In particular, the building of public environmental infrastructure utilities was regulated by a firm use of mandatory emission guidelines.

3.2.2.5 Compensatory reactions

Due to the comprehensive *Plan for Aquatic Environment* in 1987, approximately €1 billion were invested by municipalities in biological and mechanical wastewater treatment plants over a 10-year period. This was in order to fulfil the general mandatory guidelines for organic effluents (on Nitrogen, Phosphate and B15) that followed the action plan. The Plan for Aquatic Environment was related to problems of regional and national eutrophication in the mid-1980s, stemming from diffusion of organic substances from industrial point sources, agriculture, and public sewage treatment plants. Also all larger industries with direct outlets to fjords and bays were faced with new emission standards. Smaller industries were obliged to link up to, and sign contracts with, public wastewater treatment plants, and pay tax related to the contracted amount and type of discharges. The capacity of the wastewater treatment plants thus became an intermediate variable for the corporate individual standards on organic substances. Also, other pollutant-handling infrastructures were faced with standards on emissions and depositing, which were passed on to their subcontractors. The increased amount of waste handled by incineration plants led to mandatory emission standards. Instead of solely recipient assessments, the emission standards placed a focus on toxicity, exposure, diffusion, and quantities.

Through these emission standards, and the establishment of pollutant-handling infrastructure, industry learnt that there were *external solutions* to their waste problems, and if these were used or installed, no liability could be imposed on the polluter. Environmental protection was still an add-on cost, also for the public through tax-funded municipal treatment plants and subsidised industry. The waste-handling infrastructure produced *problem shifting,* as new media suffered from the pollution from incineration, wastewater sludge and recycled composites. Public authorities found that the strategy was costly in terms of management, and that they had to install new equipment for handling wastewater sludge, and solid waste. Industry learnt from the authorities' difficulties in enforcing permits and licenses and enhancing control measures, that it was industry that could process valuable, but inaccessible, knowledge concerning very complicated compositions of chemicals, the shifts in intensity of production, knowledge of control, and measuring equipment's quality. Also, under such regimes, there were no

---

[6]   In total, approximately 700 million ECU were spent until 1986 when the scheme ended.

legal or institutional mechanisms to enhance the dynamic development process of corporate capacity to continuously solve environmental problems. Finally, the same problem occurred as with the ambient quality strategy: there was no integrated or holistic approach, no focus on the core of the technology used in production, no focus on resource inputs.

Under pressure from a growing public concern about pollution at the beginning of the 1980s, the Danish liberal-conservative minority governments in 1982-1992 were pressed by the so-called green majority in parliament to be open to claims from environmental NGOs over lax enforcement of environmental protection. Thus began a period of various enforcement efforts and initiatives were sought for less costly diversified regulatory instruments. At the same time, a neo-liberal economic discourse flawed the debate and political discussions were on modernising the state. For environmental policy, this meant debates and strategies for more efficient deregulation with the use of market-based instruments, eco-labelling, and negotiated agreements. Ideologically, but also backed by environmental economic theory, state bureaucratic enforcement was questioned as the way forward for future regulation. Instead, measures to reinternalise environmental concern by economic incentives were favoured, and the discursive constellation of regulatory ideas beyond legal rules began. Although given high political attention, the liberal governments of 1982-1992 introduced very few ecological taxes. In this period the discourse on *ecological modernisation, cleaner technology, and pollution prevention* was favoured as a means to reconcile the economy and the environment through 'win-win-situations'. Some strategic policy documents were published concerning economic profits from exporting environmental-oriented systems and the competitive advantages of Danish EP, claiming to encourage firms to be technologically innovative.[7] The growing demands for improved environmental infrastructure were strategically met by efforts to link Danish subcontractors to the development. Business consultants, the environmental infrastructure industry, and the Ministry of Industry were active in the specific *business-regulation consultation forum* (Formidlingsrådet) established in 1980. Some of the softer policy measures, such as negotiated agreements, eco-labelling, green accounting schemes, and environmental management were imported from abroad in the late 1980s, but did not become full practices until the 1990s.

Both the efficiency and enforcement discussions led to various rationalisations and enforcement initiatives concerning permitting and controlling firms. Centralisation, moving the administrative issuing of

---

[7] *Erhvervsudvikling og Miljø*, Industriministeriet 1987, *Enkelt og Effektivt*, Miljøministeriet 1988, *Betænkning om miljøreguleringer og økonomisk vækst fra Metal's miljøudvalg*, København 1988.

permits and controlling activities from municipalities to counties, was initiated in order to improve politically-independent enforcement. Further a step-by-step reduction in the number of firms obliged to apply for approval (initially from about 40,000 to 25,000) was quietly started, but local pollution controls were also granted more resources.

3.2.3   New developments

3.2.3.1   Modernisation and internalisation

When the social democrats (Socialdemokratiet) took over government in 1992 they enacted former ideologically-prepared paradigmatic changes in environmental policy. This concerned advancing both supply and demand side policies for fostering pollution prevention and technological innovation. Accordingly, a number of new initiatives were seen: covenants; green taxes on car gases, CO_2, and wastewater; environmental management schemes; and green public procurement. The cleaner technology strategy in industrial environmental regulation, described above, was set as an ecological modernisation strategy. The subsidy programmes were backed up by new demand-side policies such as supporting industrialists in lobbying for green technical standards, initiating public green purchasing, supporting networking and marketing for export of Danish environmental technology, working for the better use of eco-labelling by R&D, and introducing the Scandinavian label 'The Swan'. Further, a number of specific action plans and programmes have been decided upon, where a strategic effort is sought to phase out toxic substances, to enhance organic food products, to recycle waste, to reduce car traffic, and to secure drinking water resources.

During the 1980s and 1990s a new type of environmental policy has been introduced that can be characterised as cumulative, chain-oriented, resource management. In reality it is not a deliberate strategic effort, but the accumulated results of incremental changes and new planning fields that are being evaluated and responded to. To some extent, this policy style has had effects on industry. If we look at the 1987 Plan for Aquatic Environment, we see it was paralleled with an Action Plan for Pesticides. These two plans have initiated a number of follow-up plans and orders to ensure the better effectiveness of regulations towards quality of drinking water and coastal waters. These cover: new effluent discharge thresholds for industrial waste water; phasing out of a number of pesticides and taxation on their use; demands on phasing out heavy metals for industry discharging to public sewers; health-related thresholds for toxic substances from waste water sludge; enlarged areas of wetlands; pesticide-free zones and drinking water priority zones; plans for sustainable agriculture; and the phasing out of pesticides.

The extended pollution concept and a preventive focus were introduced with the new *Environment Protection Act* of 1991. The law, basically still current, aimed at changing industry's use of manufacturing processes, raw materials, additives and products: in order to reduce the amount of resources used, and the output of pollution and waste. As a new act it had a focus on Best Available Technology (BAT) and Life Cycle Assessment (LCA) perspectives, i.e. voluntary and market based instruments. A more positive attitude towards industry and its self-regulatory capacity was seen, whereas farmers received greater attention through regulatory enforcement (Miljø- og Energiministeriet, 1991). Both target groups were influenced by a general *paradigmatic shift* in the public debate and in environmental regulation: they were to become active partners in the game, more-or-less committed to participate in pollution prevention. Finally, focus on *international subjects* grew radically: on the depletion of the ozone layer, on greenhouse effects and on long term impacts of Danish consumption. Especially Eastern European pollution problems have been a focus since 1990. Consequently, new environmental aid schemes for Eastern Europe and third world countries, were started, deliberately providing new export opportunities for Danish consultants and environmental equipment industries (Ringius et al., 1996).

3.2.3.2   Deliberate focus on technology
Programmes for enhancing corporate clean technology
The *first genuine technology innovation effort* was initiated by DEPA's 1986 R&D programme for subsidising research and development that aimed at advancing Best Available Technologies (Renere teknologi programmet). The focus was often on local, but also on *national,* environmental problems. The cleaner technology programmes (CT programmes) have served to foster innovations among firms and Danish technological R&D institutions (see the next section), and to demonstrate to authorities and industries the available options for pollution prevention. The programme has been administered within a corporate management regime, where industries and R&D institutions could forward their own preferences. A vast number of reports have been produced for various trades and relating to various environmental and technological foci. In the first generation of CT programmes, the focus was on a policy for R&D in a *push for new technological inventions* to reduce hazardous emissions and waste from industrial processes. In the first, and also in the second, CT programme (1988-1992) the focus was on *single* firms. Very often the diffusion of the developed, available, technologies was, in the 1986-1992 period, yet to take place. The second generation of programmes in the 1990s was focussed on generating an in-house *corporate pull for cleaner technologies*, through initiating environmental management,

green accounting and competence building among management staff (Miljøstyrelsen, 1997).

Lately, the cleaner technology programme has shifted its *strategic technology focus from processes to products within a LCA perspective*. The areas receiving support include innovative research in technology and product design, networking and benchmarking activities for product standards, and competence building. Thus, the non-legislative demand-side policy has been continued. Following the evaluation of the CT programmes a more strategic and market-oriented approach has been introduced by DEPA: *A reinforced product oriented environmental effort* (Miljøstyrelsen, 1996). Environmental improvements in product life cycles are the stated goal of the programme. An integration of this goal in the market is seen as a prerequisite and, in general terms, the strategic approach of the programme is to bring forward institutional preconditions for such integration. A major element of the programme is the formation of three product panels (electronics, textiles, and goods transport) designated to support a product-oriented strategy looking at design, marketing and environmental communication as main strategic elements (Miljøstyrelsen, 1999). This programme marks a changed approach. Technology is no longer considered as the main barrier to environmental improvements. The barriers are located in the conduct of the players in the industry, and the way the environment have been perceived and communicated in the product chain. In its main elements the product-oriented programme changes the institutional framework. The programme focuses on the development of new competencies both at the systemic level and at the individual enterprise:

– The development of chain- and market-oriented competencies. Programme projects now address such issues as barriers to implementing eco-labelling, environmental supply management and economic assessment of the production, and purchase and sale of environmentally friendly products.

– The development of instruments for environmental documentation (Life Cycle Assessments and risk assessments, - dedicated to industries and product groups) and the establishment of a knowledge centre on the development and dissemination of knowledge on environmental documentation and communication.

– The challenge of the individual firm is seen as a move from environmental management focused on local eco-efficiency to LCA-management - involving the need to achieve new competencies on interaction with actors in the product chain and stakeholders.

BAT regulation

With the new environmental protection act, Denmark introduced, more formerly, the German and American mode of regulation, internationally known as the *technology-based performance standard* (complementary to the recipient-oriented standard). Technology based standards focus on "... *the level of pollution reduction achievable by the use of BAT"* (Miljøministeriet 1993, §2). In Denmark, the administrative staff in local authorities (such as managers in industrial sectors) was supposed to be supported in the fulfilment of this by *Branch Orientations,* and by new guidelines concerning cleaner technology. However, after a long period of slow negotiations and little success, this process was in fact given up. Thus, Denmark still lacks legislative measures to implement BAT-based environmental improvements in industry. On the other hand, quite a number of non-formal networks, campaigns, and local strategic efforts have taken place during the 1990s, where *local* environmental officers from the public administration have created their own ideas and methods to enhance a technological approach in regulations (Miljøstyrelsen, 1998). In 2000, Denmark has implemented the EU Integrated Pollution Prevention and Control (IPPC) directive that urges for greater effort into using the BAT development in regulations.[8] After six years with enacted options to use technology-based standards, the experience until now is that authorities seem lax in demanding performance figures from industry any tighter than the earlier mentioned emission and recipient guidelines. DEPA has confirmed this interpretation to the local authorities, when they have asked for agreement to use the BAT principle. Furthermore, there is a lack of any real options, in either the IPPC directive or in Danish legislation, to focus on the life cycle of products and the cumulative impacts from the total emissions and consumption of raw materials.

A paradigmatic shift in the legalistic basis in environmental policy has occurred. In accordance with a wish to diminish the list of firms needing obligatory environmental approvals, and to distinguish industries with few environmental problems, efforts were made to *deliberately differentiate among companies as to front runners, compliance-oriented, and reluctant ones*. A command-and-control line was favoured for the latter, whereas service and routine controls were prescribed for the compliance-oriented ones. But it was among the front runners, or 'wannabees', that central, local, and regional authorities have met new challenges in opting for a corporate management regime. These firms have served as sparring partners for R&D,

---

[8]   As a result the insights for the public over permit conditions will be enhanced; worker participation in environmental decisions will be improved, and re-evaluation of permits will occur more often.

for lobbying on product standards in CERN/ISO, and for enforcing environmental regulation on competitors.

According to a central report (LO 1999), the DEPA deliberately chooses to subsidise standardisation, public procurement, and technological innovation for product and process areas that are about to be included in EU legislation. In the energy field, the 1997 CO_2 tax scheme boosted the finances for subventions in a number of new technology-push areas: R&D and investments in more energy efficiency process technology, in alternative insulation materials, and in biological gas technology. The subsidy schemes have been enlarged to cover new green jobs and green public initiatives in relation to Local Agenda 21. In total, the subsidy schemes have grown to the extent that Denmark has 24 environmental subsidy schemes (Miljøstyrelsen 1999, LO 1999) covering specific and general technological areas.

3.2.3.3   Market instruments

The 1990s was a period in which a number of new policy tools related to industry were put into practice: approximately 15 environmental agreements; 25 types of eco-tax including the very important CO_2 tax; a green accounting scheme for firms; and the building of EMAS servicing institutions. Generally, a major focus on the role of green consumerism as a motor for changing markets was given high priority. These *market-pull*-oriented initiatives, which also covered the product programme, reflected an effort to initiate market dynamics where green consumerism and green purchasing by public institutions and corporate leaders were supposed to be an efficient alternative to command-and-control. In relation to this strategy, a remarkable shift in focus has occurred, to one giving the responsibility to consumers, citizens, and car drivers to act and consume in more environmentally friendly ways. The product-orientation of Danish environmental policy thereby stresses the enhancement of environmental communication and profiling between industry and consumers. The Ministry of Environment has put a lot of effort into state-guarantied eco-labels, and has also given high priority to products for consumers such as organic food and energy saving household equipment.

3.2.3.4   Action plans, inter-policy and a new role for master spatial planning

A new tendency in the 1990s was to produce a growing number of comprehensive action plans for policy sectors. Overall action plans have been produced for sustainable agriculture, transport, and energy. In addition, the Ministry of Environment has initiated a new strategic four-year environmental plan regarding the overall prioritising of initiatives, and the implementation of the Rio documents on sustainability (Miljøministeriet,

1995; 1999). Specific target areas have been promoted: pesticides, solid waste, drinking water, chemicals, chemical-polluted sites, and greenhouse impacts. Whereas the overall action plans tend to wither away, the target area plans often contain binding obligations and time scales for phasing out hazardous substances and for recycling. A national, voluntary, promotion of Local Agenda 21 has also obtained a certain central focus and support, but has yet to involve industry (Holm, 1999).

In 1992, the Minister of Environment published *Denmark Heads for Year 2018*, as the annual National Planning Document - a binding framework for regional and municipal spatial planning. Even though it was very general and full of hot air, it was visionary with a profound plea for Denmark to be an environmental pioneer and to be the first sustainable country in the world (Miljøministeren, 1992). The use of the master spatial planning system for environmental purposes was strengthened in 1995, when Denmark formally implemented the EU directive on Environmental Impact Assessment (EIA). This covered larger projects such as road construction, gas storage, power plants, and some other high-risk plants. The EIA procedures were divided between the licensing system and the spatial planning system, placing stronger demands on public hearings and documentation. An accepted EIA project gives a permit to the entrepreneur, and the project has to be included as an attachment to the regional planning document. Recently a shift in the spatial planning act was made (Miljø- og Energiministeriet, 1998), so that the authorities in spatial planning documents have to: 1) describe their activities for sustainability, 2) report forthcoming Local Agenda 21 activities that will be supported, and 3) show how public participation and cross-sector initiatives towards sustainability will be fostered. The counties and municipalities have to report, in greater detail, about what will be done to: a) cut resources in consumption and pollution, b) lower CO_2 emissions, c) enhance biological diversity. This looks like a transition in policy style, away from the voluntary approach to a firm and formal implementation process. Overall, we may interpret the changes as an incremental process of going into the fourth stage of an environment-and-development policy (Lafferty and Eckerberg, 1998: 238): the UNCED policy where the international commitments for definite contributions to global sustainability overrule the national agenda.

The strengthened position of environmental discourses in the Danish parliament since the mid-1980s has also meant a strengthened position for the Ministry of Environment in the conservative-liberal governments of 1984-1992. The media coverage and political activities following the 1987 Brundtland report provided an internationally sanctioned, discursive backup to the Minister of Environment. She succeeded in her efforts to make the government sign a strategic environmental vision for Denmark: the Action plan for Environment and Development (*Regeringens Handlingsplan for Miljø of Udvikling,* Miljøministeriet, 1988). It included an attempt to include all public sectors in a joint commitment towards sustainability. This led to new inter-policy efforts to push those ministries having responsibilities for the degradation of the environment to initiate environmental self-observation by asking for comprehensive sector action plans for their contribution to sustainable development. A positive image was created for the previously mentioned comprehensive Plan for the Aquatic Environment from 1987 based on abatement of pollution of fjords and oceans from emissions of organic compounds. The outcomes of this major political step were *firstly* the action plans for sustainable traffic (1990), energy (1990), agriculture (1991), and forestry (1994); with more or less profound aims, targets and measures. *Secondly,* a national campaign, *Our Common Future,* was started, which addressed cross-sector and participatory dimensions of environmental concern in line with the Brundtland report's recommendations. Along with the campaign, a temporary fund was administered by a partly ministerial, partly NGO, committee. The fund helped to start urban ecological initiatives, and also initiated the formalised institutionalisation of former grassroot initiatives such as the Green Information Centre. This was launched as a centre for gathering and diffusing knowledge for the public on green lifestyles, green products, and green advisers on consumption. As can be seen from the above, the Ministers of the Environment were, in the long period with a green majority in parliament, given wide room for manoeuvre: to initiate programmes and produce political strategy documents that covered far more comprehensive initiatives than the normal environmental policy. The Ministry of Foreign Affairs was accordingly pressed to share the sovereignty of aid affairs with the Ministry of Environment, since the latter headed the formulations for new environmental aid schemes for Eastern Europe and the Third World. In 1988-1993 the Minister of Environment was also charged with many administrative preparatory functions and competencies to negotiate EU legislation.

3.2.4   Evaluation of Danish EP

Danish EP that is of interest to ETP integration covers:
- spatial planning, wastewater permits and licenses, and permits on pollution;
- norms and standards for industrial processes, agriculture, products, and green accounting schemes;
- subventions for R&D and diffusion on cleaner technologies/products;
- subventions for environmental management systems among SMEs;
- rules for waste delivery and handling;
- voluntary agreements on PVC, package, tyres, and batteries;

– action plans for specific targets, such as pesticides and organic farming.

The Danish environmental regulation of industries has been dominated by a legislative control policy, based on use of standards. Various standards for industry's performance have been related to ambient recipient quality, to emissions, to capacity of purification treatment-plants, or to market options in technological performance. Today, all these standards co-exist and thus represent very different concepts of the problem, causal mechanisms, and responses to pollution. Some existing standards have been developed along side rules from the 1970s and early 1980s, to provide industry and agriculture with legal property rights for continued pollution and expansion. Other rules from the 1980s and 1990s prescribe actors with innovative responsibility. The multiple standards are said to be confusing to manufacturers and farmers. The rules also cover so many specific media, substances, and technologies that problems are passed from one medium to another, e.g. wastewater treatment creates problems for the dilution of sludge on crop fields. The displacement of pollution from one media to another is thus seldom resolved, as standards and regulation are directed towards the distinct outlets of waste, emissions to air, and wastewater discharges.

Table 3.1: Model of cumulative expansions in the environmental regulatory focus on industry: strategy in relation to measures of planning and control. Inspired by J.P. Mortensen (1999)

Strategic aim	Separation of interests	Building infrastructure for pollutants	Incorporating environmental concern into business dynamics	Product stewardship
Response strategy	Diffusion of industry and pollutants	Transforming pollutants	Recycling	Production chain stewardship and downscaling
Means of planning	Spatial and ambient environmental quality planning	Potential impact assessment, upstream co-ordination, contracts	BAT – negotiations and technology surveys, R&D	R&D of product design, public procurement
Means of control	Environmental quality monitoring AEQ	Emissions per outlet	Waste and spill efficiency	Input of resources state eco-labelling
Involved levels of business organisation	None		Environmental managers	Product designers, value chain stewardship
Technological innovation	None	Environmental equipment for incineration, wastewater etc	New management tools, non-low waste technology in processes	De-materialising, recycling, new products, organic food,

The most obvious strength of Danish EP is that it is developing in quite a responsive way, enabling the Ministry of Environment to take up new environmental problems and initiatives from research results, public debates, and business technology options at very short warning. The consensus policy style is, according to general evaluation studies of Danish EP (M.S. Andersen, 1995; Andersen and Liefferink, 1997; Christiansen, 1996; Holm, 1999; Mortensen, 1999; Nielsen, 1997; Wallace, 1995), both a strength and a weakness. A strength because industrial trades are open to communications on technological options and to participating in finding environmental protection measures they may benefit from. There is a practical capacity for co-ordinating, consulting, and environmental monitoring with R&D initiatives in cleaner substitutes or technologies, which have established effective feedback mechanisms between the DEPA, the political system, research, and networking. The weakness is that the minister and the DEPA very often hesitate to enforce legal and binding rules that industry strongly opposes. A further weakness is the risks of falling into technological trajectories that fit whatever industrial suppliers of environmental equipment offer. A strength of Danish environmental policy is that it continuously succeeds in keeping an inter-policy perspective on the agenda, whereby the ministerial sectors for transport, agriculture, and foreign affairs incorporate strategic environmental concerns in their programmes and planning. Experts disagree as to whether standard-based environmental regulation has initiated good environmental performance among Danish industries, or whether it is market and consumer conditions that are the driving forces (Nielsen, 1997; Andersen and Jørgensen, 1998; Christiansen, 1996; Wallace, 1995).

According to M.S. Andersen (1995), it is a weakness of the basic environmental policy that the polluter-pays principle has not been taken seriously in Denmark, passing the costs for the environmental infrastructure onto the taxpayer. As a result, incentives for technological innovation have been too scarce. The discussion and debate on the effectiveness and efficiency of Danish EP have been intense since the mid 1980s and both neo-liberals and greens have had successes in their efforts to influence regulation. New market-based instruments have incrementally been established as stable parts of the tool box, efficiency measures and cuts in regulation and control have prevailed, while a broad range of action plans, bans and precautionary steps in EP have satisfied the greens.

3.3   Environment-oriented technology policy in Denmark

As described above, environmental policy is, by and large, positioned within one Ministry and a single sector political-administrative unit among the regional and municipal institutions. Technology policy is more dispersed as a policy field because technology is seldom regarded as a goal in itself, even though the 1980s in Denmark did show some tendencies in that direction. We may define technology policy[9] as efforts among public authorities (possibly in co-operation with NGOs) to deliberately enhance, moderate and diffuse: 1) a certain type of technology (e.g. biotechnology in food production); and 2) a certain development path for various kinds of technologies to fulfil certain political goals (e.g. enhancing R&D in developing water based paint products). Many ministries have hosted the dispersed technology policy - Trade and Industry, Towns and Housing, Transportation, Food, Health, and Environment - and also the implementation of technology policy: R&D schemes, action programmes and policies aiming for certain technological development paths. To track the ETP of various sector efforts is thus a somewhat difficult affair, as one has to search for a variety of non-interrelated, deliberate efforts to enhance environmentally friendly technology development within each policy sector. One way to trace ETP would be to follow specific key features, infrastructures, or themes of importance, where environmental benefits are targeted - e.g. technologies for organic farming, for substituting for detergents, for recycling electronic equipment. Many interpolicy institutional arrangements between EP and ETP will occur, and one would have to find a pattern or discursive stabilisation among the variety of efforts. Here we can take our starting point in the main TP ministry in this respect, the Ministry of Trade and Industry (MTI), and ask them how they present their networking with the Ministry of Environment and Energy (MEE) in cross-sector co-operation and ETP programmes for the benefit of the environment. This enables a first-hand presentation of deliberate ETP.

3.3.1   Institutional characteristics

Since the Ministry of Trade and Industry (MTI) was established in 1908 the name, the responsibilities, and the organisation of the ministry has been changed several times. In relation to ETP, the organisation of the ministry as it is today is best described as follows (LO, 1999: 124-131; S. Worm, MTI, interview April 1999). Firstly, the MTI is the most important actor in the elaboration and implementation of industrial policies. The ministry's department has four centres that elaborate the development of new policies. The most important centre, in an ETP perspective, is the Centre of Trade Policies. The centre is responsible for R&D, innovation, technology, and SME initiatives, and in 2000 was running two strategic ETP projects: *a*

---

[9]   By this definition, *unintended technological spin-offs or innovative impacts* from policies, e.g. EP, is not the focus.

Green Trade Policy, and a project on *Integration of environment and sustainable development in EU's internal market and industrial policy.* The ministry contains six main authorities, agencies or offices, of which the Danish Agency for Trade and Industry (DATI) administers the business-oriented industrial policies and funds granted to the network of private Approved Technological Institutes (GTSs) and the regional Technological Information Centres (TICs). The National Consumer Agency is also important, launching campaigns and control schemes for product quality, and the Patent and Trademark Office procures knowledge about industrial property rights to research and development.

The regional level is organised in a different way in terms of EP. Some counties and municipalities started regional and local initiatives in order to strengthen their industrial development from the beginning of the 1980s. A new law in 1992 improved the possibilities for a local authority to participate in such initiatives. Generally, the institutional setting of policies is organised with a lot of actors from the local authorities, firms, and elsewhere organised in local or regional Industrial Development Councils. In 1997, the government made it possible for the regional actors in industrial policies to receive support for the establishment of regional 'Business Cross-point' centres in order to strengthen the local conditions for the industry. At present there are 11 approved Regional Business Cross-point Centres (LO, 1999: 120) with the participation of municipalities, counties, Regional Labour Market Councils, Regional Industrial Development Councils, TICs, branch organisations, and Labour Market Organisations.

The development of ETP is best understood as inter-linked changes on three levels:

– The ministry's department and its Centre of Trade Policies are charged with the strategic affairs. They consider, at a general level, how for example industrial development and environment can be reconciled. In addition, the ministry is involved in reflective practices, considering the future development of its responsibilities and jurisdiction, and its relationships to other ministries and non-governmental actors. Shifting conceptions of 'the environment' prevail in its policy papers. For example, in some cases environmental concern and regulation is understood as a threat to the competitiveness of Danish firms. At other times environmental awareness is seen as new opportunities.

– The second level is the Danish Agency for Trade and Industry (DATI) that offers companies instruments to improve competitiveness and innovation processes. The agency is responsible for the programmes decided by Parliament, and has the overall responsibility for the maintenance and the development of the technological service network.

–   The third level is the specific programmes and the technological service
    network that directly interact with firms in innovation processes and the
    development of competencies. The network is the most important way to
    channel the funds that the state uses to support business-oriented
    innovation processes.

3.3.1.1   The absence of an environmental orientation in the standard TP
*   system in Denmark*

Generally speaking, there is no standard ETP system in Denmark.
Technological orientation was included as a part of the environmental policy
system in the mid-1980s when the 'cleaner technology' programme started.
Therefore, the purpose of this section is to describe the environment-oriented
technology policies as a part of the technology policy system in Denmark. In
this connection, technology policy will be understood as state actions whose
main purpose is to exert influence on the development or the use of
technology in society.

    According to Peter Munk Christiansen (1988), the development of
Danish technology policy[10] has traditionally reflected a liberal paradigm of
indirect interventions into the technological processes of the market. The
means and instruments have been general, indirect, and passive. Until the
middle of the 1980s this policy was considered to be in close accordance
with the structural traits of Danish industry. From 1984-1985, this view of
the structural character of the industry and the role of technology policy
gradually changed. It was stated that Danish industry had severe structural
problems and that the industry had to be strengthened and restructured. As a
consequence, the technology policy was redefined.

3.3.2   The standard TP system in Denmark

The first steps towards a technology policy system were taken in 1937. ATV
(Academy for the Technical Sciences) institutes (spin-off from research
institutes) were established as a co-operation project between people from
the Polytechnic School (now the Technical University of Denmark) and
certain manufacturing firms. The purpose was to strengthen industry-related
research. Since then, the ATV institutes have initiated the establishment of

---

[10] Peter Munk Christriansen's book 'Teknologi mellem stat og marked' (Technology
    between state and market' - Danish technology policies 1970-1985) was published in
    1988. The book describes the very fundamental change in Danish technology policy in the
    mid-1980s. The first part of this section of the paper is primarily based on this book
    because it tells the story of the traditional approach to technology policies in Denmark, the
    legal framework, the key actors and institutions involved, the division of labour and the
    dominant policy instruments. However, the book describes an approach that to a certain
    degree changed again around 1990.

approximately twenty research institutes. These were often related to specific branches of industry and with a very narrow scope of research. Although the ATV was still privately owned, and its membership was primarily industrial companies and academic staff from the technical university, the basic research was financed by the state (Strunge, 1993: 119).

In 1962, Parliament passed an act establishing technological institutes. The most important of these was the Technological Institute near Copenhagen and the Technological Institute in Jutland. These were merged in 1990. Their purpose was to pass on knowledge from research institutions and to participate in industrially-related R&D.

Technology policy did not become a distinct policy field until 1973, when a law regulating the larger part of technological state-market relationships was passed (Lov om teknologisk service). This law formulated a general, passive, and non-interventionist strategy. Research policy and technology policy were separated. Research policy was placed under the Ministry of Education and Research (now the Ministry of Research). The independent research institutions (ATV) and the Technological Institutes were joined under a common system known as the Technological Service Network, which became the main element of technology policy.

Until 1984, technology policy developed by chance and without an overall plan. It was organised around two institutions, the Development Fund, and the Board of Technology, and structured in four functional areas: 1) industrial research; 2) knowledge gaining and diffusion; 3) technology and product development inside the firms; and 4) building of uniform technological frameworks and development conditions – i.e. the technological infrastructure. The Board of Technology, which was placed under the technology governing body (now the Danish Agency for Trade and Industry) was a co-operative, singled out by MTI with the overall responsibility for the co-ordination of efforts in relation to the institutions of technology policies and to the firms. One of the projects concerned the environment. Another important result from the work of the Board of Technology in this period was a strengthening of technology communication with a special emphasis on the needs of SMEs, since regional technological information centres (TICs) were built up in all counties.

Between 1983 and 1986, policy objectives and instruments changed dramatically. The process of change was not guided by an overall plan but was chaotic, and the objectives and elements of the new strategy were not clear and visible until 1986. Contrary to the traditional approach, technology policy was now oriented towards *re-industrialisation and structural change*. The most important document was named: *A discussion paper on Growth and Reorganisation*, dated May 1986, and was produced in a collaboration among the Ministry of Finance, the Ministry of Trade and Industry, the

Ministry of Labour, and the Ministry of Education. It underlined that Technology Policy (TP) should secure socioeconomic growth, balance, and employment, and also that there should be co-operation between applied and fundamental research. At this stage, environmental considerations were not yet an important driver in pursuing innovation and technological development.

Resources grew, and new instruments were elaborated, mainly in the form of large research and development programmes within information technology, basic science, and biotechnology. Part of the general research policy was reformulated towards targeting industrial development. Environmental improvements were among the targeted aims of these efforts. Despite a comprehensive reformulation of the strategy, the technology policy only contributed marginally to re-industrialisation and to changing the structural problems of industry. This was mainly due to the fact that most resources were used to 'self-service' schemes, and there was very little power for structural change (Christiansen, 1988: 331).

Therefore, the main result of this phase was a growing intertwinement between state and market due to the publicly financed part of R&D expenditure in industry doubling and amounting to one fifth of the total. Furthermore, the programmes strengthened the relationships between the technological infrastructure and private firms. The state financed R&D programmes in private firms, firms financed R&D projects in public research institutions, and a lot of projects were collaborations. Although the technology development strategy was changed dramatically in the late-1980s as a result of the lack of success and for financial reasons, the intertwining between private firms and the technological service network has sustained. This was also shown in the new trade and industry strategies, where there was a focus on the establishment of networks (the end of the 1980s) and the establishment of 'resource areas' (from around 1992), where there was attached importance to industrial clusters.

It is possible to summarise the traditional TP system as follows:
- Its purpose has been to support the actors in the market in technology development and use. The idea has been to hasten development and implementation. There have been only very few examples of directly identifying certain technologies as preferable to develop. There has been programmes oriented towards specific areas, for example the technological development programme (directed towards information technology), the bio-technological programme, and the material technological programmes. However, these programmes were not directed towards the development of specific generic technologies. The main focus was to make these kinds of technologies accessible by Danish firms.

- The corporate/negotiated political system has been crucial in steering the TP system. Organisations have played a crucial role and there has been a high level of consensus about the system.
- From the mid-1980s, the technology policy became an important part of the socioeconomic discourse, i.e. technology is seen as a decisive factor in relation to employment and to the development of the competitiveness of industry, including the structural changeover of the industry. The programmes that were carried through did, however, not promote such a development, due to the fact that they mostly consisted of self-service arrangements, in which the existing technology was supported.
- The subsidy element has traditionally played a minor role. Technology policy concentrated on the building up of a technological service network, which to a larger degree was corporately controlled. The programmes from the mid-1980s have had a higher level of subsidies, but also here the effort to integrate the firms and the technological service network was emphasised. An important outcome of the shift in technology policy in the mid-1980s was that a higher degree of integration between the R&D in firms, and the R&D in the technologic service network, was generated.

3.3.3   New developments in environment-oriented technology policy

3.3.3.1   At the level of the Ministry of Trade and Industry
From around 1986, the eco-modernist discourse gradually became an element of the strategic thinking in the administration, especially in the Ministry of Environment and the Danish Environmental Protection Agency. Earlier there had been some discussions about the possibility of gaining competitive advantages in the production of environmental protection equipment but generally 'environmental protection' was understood as a necessary cost resulting in higher overall production costs. In the second half of the 1980s the MTI pointed out that the production of clean technology was an important branch of industry where competitive advantages ('first mover advantages') could be established as a result of environmental regulation. This perspective was still present in the analyses of resource areas for Danish industry performed in 1994, in which 'the environment and energy cluster' was identified as very important. The studies of resource areas resulted in several industrial policy initiatives in order to strengthen the competitiveness of the sectors (resource areas) and the innovative capacity of the firms (Christiansen, 1994). In 1986, it was stated that environmental demands could support innovation processes in the manufacturing industry (Industriministeriet, 1987; Arbejdsministeriet, 1986). However, this

perspective was only an integrated and important part of the technology policy in a short period at the beginning of the 1990s.

In 1992, the report *'Environment and Industry'* (Industriministeriet og Miljøministeriet, 1992) was published as a result of a collaboration between the MTI and the Ministry of Environment. The report stated that the two ministries have different (but not conflicting) responsibilities in relation to industry. Using 'sustainable development' as a central concept they pointed to the 'types of industrialisation that gives economic, ecological, and social advantages for existing and future generations without damaging the fundamental ecological processes'. This concept established a *common frame of understanding*. From an industrial policy perspective, this points to the fact that it has to be considered how environmental concern could be integrated into the subsidy schemes, and how the normative surplus in the environmental policies could be made a factor in securing industrial growth. It was concerned with how environmental demands could be issued in order to establish a basis for an environmental equipment industry and for new markets where environmental policies could give competitive advantages, for example by supplying Danish industry with a green image (M.S. Andersen, 1995: 145-46). At an institutional level, the report suggested a co-operation between the two ministries in the form of a permanent joint committee, mutual representation on relevant councils, and, for example a close co-operation in the standardisation work groups in CEN and CENELEC (European Committee for Electrotechnical Standardisation). The permanent joint committee, however, was not established, and it was only later that the strategy for mutual representation on relevant councils was resumed (S. Worm, MTI, Interview, April 1999). From 1993 onwards, interpolicy co-operation suffered from the MTI Minister's focus on the administrative burden on companies due, for example to environmental regulation (Støvring, 1999).

Since 1998, however, the programmatic and structural reformation by the new government, led by the Social Democratic Party, gave impetus to environmental interpolicy integration in a number of sectors, including trade and industry. The signals for a interpolicy change were reflected in the MTI's annual report in 1998 where an emphasis was put on the major competitive advantages of flexible businesses capable of adjusting to new societal demands (social, environmental) and responding in an innovatively way (Erhvervsministeriet, 1998a). The role of TP was given the task of providing consumers and business with appropriate information on environmental profiles, and a 1998 initiative, labelled the *'Environmental Challenge'*, was for making aimed at the development of a new development strategy for the integration of environmental concerns in industrial and consumer policy (Erhvervsministeriet, 1997b). New green consumer preferences played a major discursive role for the policy adjustment in the MTI. Accordingly, eco-labels such as the European Flower and the Nordic Swan began to be seen as very important for the integration of environmental concerns in industrial development (S. Worm, MTI, Interview April 1999; Erhvervsministeriet, 1998a). Further, EP-TP interpolicy adjustments were favoured to ensure corporate access to, and capability of addressing available environmental technologies (technical standards, EMAS, development contracts). In order to avoid competitive disadvantages from costly environmental demands, and to enable business to approach demands innovatively, environmental regulation was to be adjusted to corporate cultures, timing, and competitiveness (Erhvervsministeriet, 1999).

These initiatives have enhanced ecological modernisation discourses at the top level in the Danish government's general business policy. Thus, in spring 2000, the government launched a new strategy for a Danish business policy, *dk21*,[11] elaborated within an inter-ministerial group of nine ministries, where the neo-liberal project was rejected in favour of mixed economic projects to de-couple economic growth from environmental pressure. A strong, but responsive, state was seen as necessary in order to favour a sustainable society, balancing competitiveness with welfare, environment, and values. The leading environmental profile of Denmark was to be strengthened: out of six success indicators, environmental profiles for resource efficiency, environmental management systems, and eco-labelled products were to be benchmarked within an OECD comparison. A green market was to be supported, environmental production chain management and communication was seen as necessary, as was product information for consumers. The means used are enhanced green public procurement, support for exports of organic goods, further green competencies in the R&D networks to companies, and green innovation initiatives in specific trades and product groups. Finally, a green business strategy was currently being developed (mid-2000) resulting in concrete initiatives.

3.3.3.2 *The technological service network - building of environmental competencies*

The service network structure was changed in 1996, as a new law was passed about technological service, and a *new 'Council for Technological Service'* was made responsible for developing the overall strategy for Approved Technological Service Centres (GTSs). In this council, there were personally-appointed people representing NGOs and public authorities (including one representing the DEPA). At the end of the 1980s, the technological service network was heavily dependent on funds from the technological development programmes and basic funds from the Danish

---

[11] Erhvervsministeriet: dk21, København 2000 - http://www.em.dk/dk21.

Agency for Trade and Industry. Therefore, the reduction in funding for technological politics at the beginning of the 1990s made the network very dependent on their commercial contracts. In order to maintain the competencies in the network, it was decided to increase the basic funds to the GTSs so that, in 1997, the total amount of basic funding (contract funds) was €37.8 million out of a total turnover of €258 million. Furthermore, it was decided to concentrate the effort. There were, in 2000, only fourteen certified centres, some of which offer a wide range of services (for example the Danish Technological Institute) while others, in the ATV tradition, are very specialised (for example the Danish Institute of Fire Technology).

The governance of the technological service network has changed. The basic funds were to be related to '*Result-dependent contracts*' (offered for three years), and the Council for Technical Services was made responsible for the co-ordination between the Institutes. It is seen as very important that the centre addresses SMEs, either directly or through the TICs in order to support the diffusion and the adoption of new technologies. An evaluation of the users' satisfaction with technological services concluded that 33% found the topic 'environment and energy technology' to be the most important area (Rådet for teknologisk service, 1998). A specific evaluation of the TICs has also been undertaken, but this does not evaluate the centres' capability of supporting environmental development. (Erhvervsfremmestyrelsen, 1999).

Six of the fourteen GTSs are engaged in ETP-related activities. These are the Danish Hydraulic Institute, the Danish Standard Organisation, the Danish Technological Institute, the Danish Toxicology Centre, the Danish Water Quality Centre, and Dk-teknik Energy and Environment. From an environmental perspective, DTI is by far the most important. DTI-Environment has nine different departments (such as bio technology, chemical technology, life cycle analysis, industrial environmental technology and waste technology). Another important activity is DTI-Energy.

3.3.3.3   New environment-oriented technology policy instruments

One of the first large ETP programmes was the '*Industrial use of environmental technology*'. The Danish Agency for Trade and Industry (DATI) started the programme in 1991, with the purpose of promoting the export of Danish environmental technology (the total amount of money used in the programme was €15 million). The programme was part of an extensive initiative by the Environmental Protection Agency to establish a co-ordinated effort in the environmental area. DEPA's part of the project was ETP-oriented, focusing on a push for the development of new technical solutions. The MTI part focused more on capacity building and market conditioning, and the efforts were split into four directions: on international contacts, on the establishment of export-oriented business networks for environmental technology, on a promotion and marketing strategy, and finally on effecting international standardisation of environmental profiles (LO, 1999: 36). The evaluation of the scheme found that it was only partly a success: new international contacts were established, and environmental profiling of standards occurred. However, environmental marketing failed, and so did networks and efforts to raise 'green sales'.

A new ETP instrument was the €38 million programme (1997-2000) for so-called '*Innovation Centres without walls*' (Erhvervsministeriet, 1997a). In these centres, personnel from universities, GTSs, private firms, research centres, and financial institutions were supposed to work together. One of these centres, the Centre for Research in Freight Transportation and Logistics, has focused on the use and implementation of new technology, the development of competencies, new collaboration forms in the transportation chain, and sustainable transportation (Erhvervsministeriet, 1998b). Another new instrument in 1995 was *Centre contracts* (€14 million in 1999). The overall purpose was to strengthen the relationships among research, technological services, and firms. According to an evaluation by DATI, there have so far been nine centres of relevance in an ETP perspective (LO, ibid.). An example in the environmental area is the Technology Centre for Information. The scheme has been evaluated (Erhvervsfremmestyrelsen, 1998b), but none of the centres has yet concluded its activities. Therefore there is no definitive conclusion, and there has been no specific evaluation of 'Centre contracts' from an ETP perspective.

Development contracts were another new instrument from 1994 onwards (€13.7 million in 1999). Their purpose was to strengthen public-private collaboration in the development of new products and services (Erhvervsfremmestyrelsen, 1995). Since 1994, 40 commercial contracts, 10 non-profit contracts, and 40 feasibility projects have been started. The scheme was evaluated in 1997 and it was concluded that the prime effect was to reduce prejudices and to establish networks between private and public actors (Erhvervsfremmestyrelsen, 1998a; 1998b). There have been several projects with environmental goals, for example a 'green' kindergarten using urban ecology principles designed by a municipality in co-operation with an architect (the goal was to combine health, ecology, and especially energy savings).

The 1995 *scheme for Environmental accounting systems* (€2 million in 1999), noted earlier, is also relevant from the perspective of inter-policy co-ordination. The scheme, administered in collaboration with the Environmental Protection Agency, was directed to branch associations and other advisors developing environmental accounting systems especially for SMEs. The scheme was supposed to make it possible for firms to receive an

EMAS certificate. Subsidies were for information activities, the development of training initiatives, materials developing, and the testing of new equipment and methods.[12] The evaluation of the schemes is ongoing, but so far the results have been evaluated positively.

Finally, in the environmental area, *'Icebreaker'* was put out as a scheme (€8.2 million in 1999). The goal has been to support the employment of experts in the environmental area and to promote activities in this area. The scheme may finance half of the wage costs of formerly unemployed people for up to nine months. The interest from firms has been rather limited, and only fifty percent of the subsidy fund from 1998 was used. There has not been an evaluation of this scheme.

3.3.4   Evaluation of the Danish ETP system

While the Danish ETP system has been an integrated part of the TP system, there are only a few schemes with an exclusively environmental orientation, such as the environmental management scheme and the subsidy schemes for renewable energy. Most of the programmes have subsidised specific institutional forms of collaboration among public and private actors, GTSs, and TICs. That is, R&D is seen as the appropriate means to ensure corporate capability to adapt cleaner technologies, made accessible through information campaigns and training. Environmental orientation has, to some degree, been integrated in all of these programmes. Furthermore, Danish TP has been focusing on strengthening the competencies in the Technological Service Network. Therefore, a deeper understanding of ETP in Denmark demands a thorough analysis of the specific interaction between the Technological Service Network and enterprises in environment-oriented innovation projects.

In the last few years, almost every part of the Danish TP system has been evaluated. These evaluations focused on the institutional set-up, and the opinions of various users of the system. There has been no specific evaluation of the system from an ETP perspective, although ETP projects are often mentioned as examples of the societal potential of the Danish systems. Stakeholders have not discussed the TP or the ETP system in public to any great extent in the 1990s. There was a lot of discussion in the late 1980s but since than the conflicts on the development of the system have been a part of the day to day business in the corporately-organised system. The system and its governance have changed - but not due to public discussions. On the contrary, it was presented as a way of making the system more effective in fulfilling politically defined goals. Since the end of the 1990s, the ecological modernisation approach has been partly integrated into the ministry's

---

[12] Environmental Protections Agency's homepage.

strategic goal orientation, where green competitiveness became to be seen as an advanced national competence globally. Therefore, various strategic and policy documents have been published, and a partial re-orientation of ETP has been scheduled, still predominantly on the path of developing corporate readiness and capability, and making the green market more visible. As a dynamic, state, impetus for green market establishment, benchmarking will be fostered through the use of strategic indicators of eco-efficiency, accounting for eco-labelled green products, and environmental management systems. New incitements for shaping a green market will be provided, as will specific green clusters. Thus, it seems to be the case that after nearly fifteen years of random and dispersed ETP initiatives, the MTI is looking for an active role in eco-modernisation discourse-coalitions, where inter-policy co-operation with MEE will seek to foster an innovation-adjusted and competitiveness-oriented EP.

3.4   Concluding remarks on the integration of ETP and EP

In this chapter we have told two different stories. *On the one hand* we have told the story of how the EP system has adapted regulatory concepts and discourses from the Keynesian and neo-liberal theories, to which the TP system, dominated by the MTI, has also adhered. Thus, firstly the accommodation of EP demands to the technical and economic options among environmental technology suppliers, and secondly the trust in R&D-based technology pushes for development and diffusion of new and more environmentally friendly technologies. This culminated in a strategic orientation towards an EP that, by proper institutional framing and strategic alliance with first movers, would generate green market dynamics wherein industry would seek to improve environmental protection through the marketing of green products. In terms of industrial pollution, technology has become the core object; and markets, consumers, and managers the core subjects. *On the other hand* we have told how TP has changed its responses to environmental demands; from a hostile position to avoid add-on costs, through supporting renewable energy and environmental infrastructure equipment niches, and ending with an eco-modernistic discourse on the competitive advantages of greening of industry. TP has supported specific technological solutions to fulfil environmental demands, and encouraged the greening of management and markets by capability-building and institutionalising market transparency. From these two different perspectives we have shown how the same policy papers, programmes, and instruments were interpreted as a part of two different policy frames. The amount of inter-policy co-ordination shown seems to be built on, even to depend on, a mutual belief in a certain eco-modernisation: market driven benchmarking and a role for the state in building capability (R&D, information), in framing a market pull by green purchasing, and enhancing eco-labelling and international standards. Alternative regulatory responses, in alternative problem foci (e.g. property rights, mass consumption culture, and material flow cycles) have been neglected.

If we take a general look at how the interlinkage or crossover between EP and TP has developed, we may identify the following stages:

1. 1972-1984: Consensus-oriented building of *guiding* environmental standards in close consultation with Danish industrial trades: balancing environmental requirements to available purification and dilution techniques. Subvention schemes were launched for industries having difficulties in complying with purification demands. The period was characterised by no specific technological *innovation* focus, and no specific environmental focus in the emerging TP. An exception though was a subsidy scheme from 1981 onwards for investments in renewable energy technologies.

2. 1985-1991: *A technology push* strategic co-orientation of TP and EP. Here the focus was on the technological capacity options for increased Danish subcontracts to the public and private purification infrastructure, caused by increasing demands for environmental protection. Strategic environmental modernisation by stressing the competitive advantages that might emerge from Danish environmental front running. Research and development TP programmes in environment-oriented topics. R&D EP programmes in cleaner technology and recycling related to processes and waste. A new, but GMO targeted specific risk and environment-oriented concern was launched with the 1986 act on Environment and Genetic Engineering, which banned all experiments with genetically modified organisms (GMO) unless they were given a dispensation.

3. 1992-2000: An institutionalisation of the technology-orientation of EP in acts and orders occurred. The export of Strategic systems became a new orientation for the support of developing Danish environmental infrastructure, institutionalised in various aid programmes under both Ministries (MEE and MTI). New R&D programmes were started on cleaner technology, environmental management (EMAS), technology diffusion, and cleaner *products*. Enhanced green public procurement, and institutionalised green mechanisms in *market pull* policies, were given priority in both EP and TP, especially for organic agriculture products. Finally, new TP and EP schemes for creating new green jobs have been initiated at in the end of the period. Lately, there has developed a general discursive environmental co-orientation within TP,

but only very few specific initiatives have been launched. Enhanced administrative co-ordination has occurred since 1998.

The specific characteristics of the institutional frameworks for integration of ETP and EP in the Danish context are as follows:

- Denmark has a relatively strong Ministry of Environment and Energy. The ministry has managed to establish an eco-modernistic discourse as a part of the general policy of the government. Strategies and support schemes for cleaner technologies have become a part of its competence. The ministry has established, and been involved in, policy networks with branch organisations, private firms and other actors in order to influence technology and business policies. Furthermore, it has had many programmes focusing on the development and diffusion of technology, and it has been an important actor in the development of the technological infrastructure and green markets.
- The Ministry of Trade and Industry has also, in recent years, become a strong ministry; and business-related research and technology policy is a part of the ministry's competence area. Business policy and technology policies have been consensus-oriented, and branch organisations and private firms have played an important role in the formulation of new policies.
- The technological infrastructure (the GTSs and the TICs) is an important result of the consensus-oriented business and technology policy. This infrastructure is the responsibility of DATI (under the jurisdiction of MTI) and it combines consultative work and business-related technology research financed by the government. A major part of the money used for EP and ETP is channelled through the infrastructure or is used to develop basic GTS competencies.
- There is a vertical division of labour among the state level, the county level and the municipalities. Counties and municipalities also have a major part of the responsibility for the implementation of EP and TP. Therefore, there can be large regional variation according to the degree to which local authorities, for example, try to enhance cleaner technologies as an important part of the regulations. Likewise, there are established regional TIC centres in order to support the development, adaptation, and diffusion of new technologies, especially in SMEs. Business policy institutions are also established at the local or regional levels in order to attract enterprises, or in order to strengthen networking in the local business community.

These institutional characteristics mean that there are different arenas for the interactions between ETP and EP. At the governmental level, the Ministry of

Environment and the Ministry of Trade and Industry have tried to integrate environment and technology from different perspectives. Both ministries have emphasised that it is not a hostile relationship. On some occasions, since the late 1980s, the Ministries have formulated a common strategy and have administered programmes for the development of environmentally-friendly technologies, for management systems, and for collaboration. However, the Ministry of Environment alone has administered the EP activities, and the ETP elements in TP have been administered by the MTI.

However, there have been interactions in the GTS system, and at the regional and local levels. Environment-oriented programmes have been very important for the GTSs and the TICs. Therefore, in the technological infrastructure, there has been a focus on the integration of ETP and EP, and this integration is particularly seen in some of the new programmes and support schemes such as the Centre Contract scheme.

It is important to note that EP is a rather strong policy area, while ETP must be seen as a weak area. In recent years, the development of new smart technologies, materials and products has been seen as the cornerstone in the development of more environmentally friendly and sustainable production. Thus, while competitiveness, and not environmental concerns, is understood as the key element in TP, the incremental greening of technologies and industry has recently had an impact on the strategic orientation of TP towards an integration of green competitiveness. Whether this will trickle down to institutional and regulatory efforts within TP is an open question. Nevertheless, on the general level, the policy sectors of EP and ETP have gone through a mutual reflection and recognition whereby a discursive and strategic interpolicy orientation has occurred.

REFERENCES

Andersen, Michael Skou (1995) *Governance by Green Taxes*. Manchester: Manchester University Press.

Andersen, Mikael Skou and Duncan Liefferink (1997) *European environmental policy - The pioneers*. Manchester: Manchester University Press.

Andersen, Michael Skou and Ulrik Jørgensen (1995) *Evaluering af indsatsen for renere teknologi 1987-1992, Orientering fra Miljøstyrelsen nr. 5*. København.

Andersen, Niels Aakerstrøm (1995) Selskabt forvaltning. København: Nyt fra Samfundsvidenskaberne.

Arbejdsministeriet, Finansministeriet, Industriministeriet og Undervisningsministeriet (1986) *Debatoplæg om vækst og omstilling. Krav til strukturpolitikken*. København.

Christiansen, Peter Munk and Lennart Lundquist (eds) (1996) Governing the Environment: politics, policy and organization in the Nordic countries. Copenhagen: Nordic Council of Ministers.

Christiansen, Peter Munk (1994) *Miljø og erhvervsudvikling*. København.

Christiansen, Peter Munk (1988) *Teknologi mellem stat og marked*. Aarhus: Politica.

Erhvervsfremmestyrelsen (1995) *Centerkontrakter Dansk innovation mellem erhvervsliv.* København: forskning og teknologisk service.

Erhvervsfremmestyrelsen (1998a) *Evaluering af udviklingskontraktordningen.* København.

Erhvervsfremmestyrelsen(1998b) *Evaluering af dansk teknologipolitik.* København.

Erhvervsfremmestyrelsen (1999) *Miljøstyring og miljørevision i danske virksomheder.* København.

Erhvervsministeriet (1997a) *Erhvervsredegørelse 1997.* København.

Erhvervsministeriet (1997b) *Dialog med Miljø/energi. Baggrundsanalyse - Offentlige-private selskaber på miljøområdet.* København.

Erhvervsministeriet (1998a) *Erhvervspolitisk center.* København.

Erhvervsministeriet (1998b) *Erhvervsredegørelse 1998.* København.

Erhvervsministeriet (1999) *Erhvervsredegøresle 1999.* København.

Holm , Jesper (1999) Status of Local Agenda 21 in Denmark. In: Bill Lafferty (ed.*)* *Implementing Local Agenda 21 in Europe.* Oslo: ProSus.

Industriministeriet (1987) *Erhvervsudvikling og Miljø.* København.

Industriministeriet og Miljøministeriet (1992) *Industri og miljø.* København.

Lafferty, Bill and Katharina Eckerberg (1998) *From Earth Summit to Local forum.* London: Earthscan.

LO (1999) *Grøn erhvervspolitik.* Analyse opgave i relation til start af LO's project. København.

Miljøministeriet (1988) *Enkelt og Effektivt.* København.

Miljøministeriet (1992) *Danmark på vej mod år 2018.* Landsplanredegørelse for Miljøministeren. København.

Miljøministeriet (1993) *Vejledning om godkendelse af listevirksomheder .* København.

Miljøministeriet (1995) *Natur- og Miljøpolitisk redegørelse.* København.

Miljøministeriet (1999) *Natur- og Miljøpolitisk redegørelse.* København.

Miljøministeriet (1988) *Regeringens Handlingsplan for Miljø of Udvikling.* København.

Miljøministeriet (1999) *Natur- og Miljøpolitisk redegørelse.* København.

Miljø- og Energiministeriet (1998) *Forslag til lov om ændring af lov om planlægning,* L134, 9.12.98, Folketinget København.

Miljø- og Energiministeriet (1999) *Natur og Miljø 1998.* København.

Miljøstyrelsen (1996) *En styreket Produktorienteret indsats.* København.

Miljøstyrelsen (1998) *Inddragelse af renere teknologi i tilsyns- og godkendelsesarbejdet.* Miljøprojekt nr 388. København.

Miljøstyrelsen (1999) *Prioriteringsplan for Program for renere produkter m.v 1999,* København.

Miljøstyrelsen (1996) *Cleaner Technology Projects in Denmark 1995,* Working Report no.19. København.

Moe, Mogens (1997) *Miljøret,* GAD. København.

Mortensen, J.P (1999) *Integration af kredsløbstankegang og forbedringspotentialer i miljøgodkendelsessystemet.* Roskilde: Teksam, Roskilde Universitetscenter.

Nielsen, Eskild Holm (1997) *Mejeriernes forebyggende miljøarbejde – et resultat af miljøreguleringen?* Institut for Samfundsudvikling og Planlægning. Aalborg: Aalborg Universitets Center.

Ringius, Lasse, Jesper Holm and Børge Klemmensen (1996) Danish Environmental Aid to Eastern Europe: Present and Future. In: R.E. Löfstedt and G. Sjöstedt (eds.) *Environmental Aid Programmes to Eastern Europe: Area Studies and Theoretical Applications.* London: Avebury.

Schroll, Henning (1997) Miljøreguleringens historie. In: Jesper Holm, Bente Kjærgård og Kaare Pedersen *Miljøregulering –tværfaglige analyser.* København: Samfundslitteratur.

Strunge, Lars (1993) *Det teknologiske udviklingsprogram - Fornyelse af dansk industri- og teknologipolitik.* Roskilde: Roskilde Universitetscenter.

Støvring, Suzette (1999) *Myter og miliarder – en analyse af konstruktionen af policyfeltet om virksomhedernes administrative byrder.* Roskilde: Roskilde Universitetscenter.

Wallace, David (1995) *Environmental Policy and Industrial Innovation.* London: Earthscan.

APPENDIX

List of abbreviations

DATI: Danish Agency for Trade and Industry

AEQ: Ambient environmental quality

ATV: Akademiet for de tekniske videnskaber (independent research institutions)

BAT: Best available technologies

CEN: European Committee for Standardisation

CERN: European Laboratory for Particle Physics

DEPA: National Environmental Protection Agency (Miljøstyrelsen)

DTI: Danish Technological Institute

EIA: Environmental Impact Assessment

EMAS: Eco-management and audit scheme

GMO: Genetically-modified organism

GTSs: Godkendte teknologiske institutter (approved technological service centres)

IPPC: Integrated Pollution Prevention and Control

ISO: International Standard Organisation

LCA: Life cycles analysis

LO: Landsorganisation (Trade Union Council)

MEE: Ministry of Environment and Energy

MTI: Ministry of Trade and Industry

NGO: Non-Governmental Organisation

SEIA: Strategic environmental impact assessment

SME: Small and medium sized enterprise

TPE: Technology oriented environmental policy

TICs: Teknologiske Informationscentre (regional technological information centres)

UNCED: UN Conference on Environmental and Development

Interviews

Arne Remmen, Miljørådet

Michael Søgaard Jørgensen, DTU

Dorte Rønnow, Miljøstyrelsen

Erik Højbjerg, Statskundskab- KBH's universitet

Kirsten Ramskov, Danmarks Naturfredningsforening

Frank Bill, industrikontoret - Miljøstyrelsen

Susanne Worm: Ministry of Trade and Industry

Ulla Ringbæk, Ministry of Environment and Energy, EPA.

Chapter 4

Environmental Policy and Environment-oriented Technology Policy in Germany

JOBST CONRAD
Center for Environmental Research Leipzig-Halle, Germany

4.1   Introduction

The purpose of this chapter is to provide a sketch of the development and configuration of environment and ecology-oriented technology policy co-ordination and co-operation in Germany on the empirical level. Furthermore, it will describe, in analytical terms, the structural features and problems involved therein, and draw some policy-oriented conclusions of significance for the ENVINNO-project (IIUW et al., 1998).

The chapter proceeds as follows. A short summary of the development of environmental policy (EP) and of environment-oriented technology policy (ETP) is given referring the reader for further information explicitly to the corresponding literature. Then the dominant actor configuration and institutional arrangements shaping the evolution and the degree of EP/ETP co-ordination are described, resulting in (varying) objectives of, interests in, and directions of influence in EP/ETP co-ordination. These features are illustrated by few typical examples. Finally, I point out the main determinants of EP/ETP co-operation, differentiate between fundamental and manageable problems of addressing EP/ETP co-ordination, and arrive at some policy conclusions concerning environment-oriented innovations.

The methods underlying this paper signify its sketchy character. The basis of the empirical and theoretical analysis of EP/ETP co-ordination in

Geerten J.I Schrama and Sabine Sedlacek (eds.) Environmental and Technology Policy in Europe.
Technological innovation and policy integration, 97-124. © 2003 Kluwer Academic Publishers. Printed
in the Netherlands.

Germany are a dozen interviews with persons in key institutions[1], some literature (see literature references), background knowledge on political theory, and my familiarity with EP and TP (technology policy) over several decades. Within these interviews organised on short notice, knowledge about EP/ETP co-ordination was mainly available for the last five years and hardly for earlier periods (1970s and 1980s).

Concerning the interviews, the specific perspective of the interviewee clearly influenced his/her (normative) assessment of EP/ETP co-ordination, apart from his/her institutional affiliation and position. This once more confirms the significance of the Thomas theorem[2], and allowed, as well as required, the author to take a more detached and balanced position in assessing development and functioning of EP/ETP co-ordination.

4.2   Development of environmental policy

4.2.1   History of German environmental policy

Environmental policy as a socially and substantially differentiated policy area has developed (in industrial societies) only since the 1960s, although social concern with nature protection and health policy can look back on a long tradition. In the 1960s air pollution laws existed at the state level, but it was not until 1969 that EP developed into a national policy area in Germany based on a comprehensive concept of environmental protection.

> *"In 1972, an amendment to the German constitution granted the federal government so-called concurrent legislative power for statutory regulations on waste management, air pollution control, noise abatement, protection from radiation, and on criminal law relating to environmental protection matters. Concurrent legislative power (i.e., power shared between the federal government and the states) means that the federal government has the right to issue detailed regulations, and that in this case, federal law supersedes state law. However, in the areas of water management, regional planning, nature conservation, and landscape preservation, the federal government is only authorised to issue so-called framework laws as a basis for detailed and specific legislation to be drawn up by the states." (Jänicke and Weidner, 1997: 137).*

---

[1]   BMBF (3), BMU, UMK, UBA (3), DLR, VDI, TÜV Rheinland, ISI (2), ZEW (1 to 2 hours); GSF, SRU, TUB (2), TAB, AfAS, ATA, WZB, Öko-Institut (ca. 15 min).

[2]   If man defines a situation as real, it will be real in all its consequences.

Reflected in a temporal learning process, EP has to deal with different types of environmental problems and effects, the related generation of socio-structural problems and impacts, resulting in a range of policy strategies, instruments, and divergent assessment of success or failure.

As a cross-sectional policy EP has had a hard stand in the past when environmental protection required substantial concessions by established interests concerned. As a consequence, EP could be characterised by considerable environmental legislation but weak and ineffective implementation and by a tendency to shift environmental problems in space, in time and to other media, for example, by building high stacks, by hazardous waste disposal facilities, or by water instead of air pollution. After EP had lost a lot of its initial momentum, the growing environmental consciousness of the public and ongoing major events and environmental disasters led to a revival and increasing relevance of environmental concerns in politics. The result was the establishment of green parties and a shift towards more effective measures, but also to short-term responses in EP more recently. Whereas originally environmental concerns were mostly pursued in separate policy arenas and institutions, they are currently penetrating established policy fields. These are frequently connected with severe political conflicts, whereas in the past established actors and institutions often could afford to simply ignore the political efforts and claims of environmentalists. The polluter-pays principle only partly holds in practice. Corrective policy orientations to avoid environmental damages dominated in the past. A preventive environmental policy (cf. Simonis, 1988) is demanded and supported by all actors publicly, but it is rarely actual policy practice yet. Overall, there has been a great deal of EP debate but little effective EP measures in the 1970s. In the 1980s, a gradual change towards a stronger and more effective EP could be observed. However, it was often not before the 1990s that this change led to substantive effects of environmental protection in most cases.

4.2.2   Institutions and approaches in environmental policy

Thus, EP is relatively well established since the 1990s. The federal government is primarily responsible for policy formulation and international co-operation, which has gained importance by the rising number of international environmental agreements and regimes. Environmental legislation is formulated by federal and state authorities, and at the level of the European Union (EU), too. Implementation and enforcement are mainly the task of the states and of the local authorities. The courts had at times (especially in the 1970s) a significant influence on program formulation and environmental policy implementation in Germany. Up to now, however, criminal law and the courts have played only a minor role in environmental matters.[3]

Germany has a broad network of well-established environmental policy institutions since the 1990s.

> *"At federal and state level there are specific authorities responsible for the formulation and implementation of environmental policy. Local authorities usually have environmental policy departments. At federal and state level, again, there are larger agencies responsible for environmental research, planning and development ... Meanwhile, most ministries, even the Ministry of Foreign Affairs and the Ministry of Defence, have a special environmental protection department." (Jänicke and Weidner, 1997: 143).*

In the 1980s one could observe a tendency towards informal agreements and package deals with industries concerned, and towards corporate expert networks with factual decision-making power by e.g. setting standards (cf. Wolf, 1988, 1992). In parallel, increasing environmental protest challenging the informally reached agreements before courts, and a growing role of environmental concerns in industrial development and decision-making, legislation, jurisdiction, and public debate could be observed as well.

Due to severe criticism the German government has begun a systematic review of the dominant command-and-control approach in recent years.

> *"On the one hand, it has examined possible ways and means of speeding up planning and industrial siting procedures, as a reaction to complaints from industry. On the other hand, it has modernised its instruments by stressing the role of information and negotiation." (Jänicke and Weidner, 1997: 139).*

In contemporary German EP a combination of hierarchical and co-operative elements can be seen, characterised as "negotiation under the shadow of hierarchy" (Scharpf, 1991).

> *"The motor of change in favour of ecological modernisation in the late 1980s and early 1990s was not so much government regulation, but a broad coalition of modernisers stimulating the self-interest of enterprises in respect of costs (energy, materials, waste, transport etc.) and new "green markets". The environmental awareness of consumers and the threat of government intervention, combined with negotiation and*

---

[3] The lower administrative courts have tended to rule in favour of environmental concerns; however the higher courts have overruled a number of these decisions and have in general tried to prevent the lower administrative courts from narrowing too far government's scope for discretion. (Jänicke and Weidner, 1997: 144f)

consultation, were important background facilitators in this process."[4]
(Jänicke and Weidner, 1997: 148).

4.2.3   Impacts of environmental policy

Meanwhile EP can point to some important improvements in environmental protection and quality[5] (cf. decrease in SO_2 emissions, phasing out of CFCs, better water quality of rivers; cf. BMU, 1990, 1994, 1998; UBA, 1997). Firstly, the regulatory infrastructure has been installed, as environmental information systems, the establishment of a considerable number of advisory and regulatory environmental bodies[6], and an extended overall legal and regulatory basis of environmental protection, including environmental liability[7] and an environmental statute-book (Umweltgesetzbuch). Then a gradual installation of market oriented policy instruments, and substantial advances in internalising environmental costs and in production-integrated environmental protection have occurred. Furthermore, Germany tackled the cleaning-up and mastering of severe environmental damages of the former GDR, and German environmental industry achieved a leading international position. Finally, Germany has recently shown a more comprehensive

---

[4]   Institutional capacity for environmental action must be related to functional requirements. *"Moderate goals, such as the dilution of air and water pollution, could be achieved with a low capacity for environmental policy. In Germany this was even possible without public pressure and with a small, additive administration. The end-of-pipe approach was implemented against a background of strong public pressure, changes in the political constellation, the development of a highly advanced monitoring technology and the establishment of a special eco-industry in tandem with a specialised eco-bureaucracy. But ... the capacity of environmental policy has proved too small to implement a strategy for sustainable development, which includes such difficult tasks as the stabilisation of land use, soil and ground water conservation, or systematic reduction of material inputs into the production process.... For a truly 'reductive' strategy regarding input in production, the coalition of ('additive') eco-bureaucracy and eco-industry cannot be seen as a driving force. It could be seen as a restrictive factor in such a strategy. Here it is the capacities of ecological modernisation within the economic system and the role of the green business sector, with their effects on the 'dirty' product chain, that have become more important."* (Jänicke and Weidner 1997: 150f).

[5]   However, improvements in environmental quality have been made easier by economic modernisation and increases in efficiency (Jänicke et al., 1992).

[6]   To be mentioned is in particular the establishment of SRU (Rat von Sachverständigen für Umweltfragen) in 1972, UBA (Umweltbundesamt) in 1974, BMU (Bundesumwelt-ministerium) in 1986, UGR (Beirat Umweltökonomische Gesamtrechnung) in 1992, TAA (Technischer Ausschuß Anlagensicherheit) in 1992, SFK (Störfallkommission) in 1992, and WBGU (Wissenschaftlicher Beirat Globale Umweltveränderungen) in 1992.

[7]   The environmental liability law, passed in 1990, was another important step towards generating alternative approaches to environmental regulation, though with still limited practical use for injured citizens and a damaged environment (Babel and Zschörnig, 1999).

perspective of EP towards sustainable development (cf. Enquete-Kommission, 1994; BMU, 1994, 1998; SRU, 1998).

However, a clear orientation and coherence of EP is still more prominent on the level of symbolic than of substantive politics.[8] Deficits in bureaucratic flexibility and in implementing environmental information systems and environmental liability remain obvious (Jänicke and Weidner, 1997; Jänicke et al., 1999; Weidner, 1992). Along with enhanced global competition and ongoing individualisation processes the general social and political climate seems to lose its favourable momentum for EP, thus undermining its further progressive development.

4.3   Development of environment-oriented technology policy

4.3.1   History of German science and technology policy

Science and technology policy (STP) has developed into a rather complex and differentiated system in Germany after the Second World War. The basic and strategic orientations of STP changed over time, as can be seen from the various federal research reports (Bundesforschungsberichte) between 1965 and 1998 (cf. BMBW, 1972; BMFT, 1988; BMBF, 1996, 1998). For instance, in a phase of planning euphoria around 1970, the 'lib-lab' government decided to see STP as an active political industrial structural policy in accordance with social needs (although these symbolic policy statements rarely corresponded to empirically observable STP practices). Later governments understood STP mainly as a device to support the (international) economic competitiveness of German industry in accordance with its primarily self-defined research needs.

First, the federal government was hesitant in assuming responsibilities for science and technology, and where it did, as in nuclear power, aerospace, and electronic data processing, its programs for supporting industrial technology were ineffective. The German innovation system till the 1980s mainly reflected the momentum of organisations that have existed for a long

---

[8] *"The environmental policy of the last two governments was unable to satisfy fully the demand for an environmental policy programme that is cohesive and transcends individual sectors; most regulations and measures are still concerned with individual environmental media. A number of more comprehensive schemes have, however, been initiated... Without doubt there has been some progress ... in individual areas of environmental policy. The development of a fundamentally new programme for ecological modernisation has not, however, been successful. In principle we have a fairly active, but still fundamentally technocratic, and to some extent merely symbolic policy of small steps."* (SRU, 1994: 32, 177).

time, have survived the period of wars and crises, and since then have grown in size. It was only in the late 1970s and 1980s that the BMFT (Bundesministerium für Forschung und Technologie; Federal Ministry for Research and Technology) began, hesitantly, to assume a role as a manager of a national innovation system. It designed programs so as to strengthen co-operation and the flow of personnel and information between different organisations within the system, and fostered new institutions such as the FhG (Fraunhofer-Gesellschaft; Fraunhofer-Society) that provided new links among different components of the system. By that time, however, the splitting up of federal responsibilities between the BMFT and the BMBW (Federal Ministry for Education and Science) had created a new barrier for policymakers to consider the system as a whole (Keck, 1993: 145,146).

Meanwhile, the perception and understanding of STP has become more differentiated for and among different (STP) political actors. Accordingly, the guidelines and orientations of STP are multidimensional ones (cf. Braun, 1997; Stucke, 1993).

The institutional arrangements of science and technology policy and administration have become more clear-cut during the 1970s and 1980s. In accordance with the BMBF (Bundesministerium für Bildung, Wissenschaft, Forschung und Technologie; Federal Ministry for education, science, research and technology), formed in 1994, one may distinguish between the DFG (Deutsche Forschungsgemeinschaft; German Research Society), the MPG (Max-Planck Gesellschaft; Max-Planck Society), which is responsible for most basic research institutes, and the FhG, which is the umbrella organisation for many applied research institutes. Secondly there are the large research centres, mainly jointly funded by federal and state governments, the research institutions belonging to the so-called blue list, again funded jointly by federal and state governments, and the many departmental federal research institutions. Finally there are various other research oriented settings such as technology parks, and, in particular, industrial research, which meanwhile spends about 70% and typically funds ca. 50% of all R&D expenditures. Military research plays a relatively minor though no marginal role in research and development and is mainly funded and coordinated by the Ministry of Defence.

All these research settings dispose of formal and informal advisory bodies. Furthermore, the BMBF or former BMFT have installed around 15 project management bodies (Projektträger), mainly within large research centres, which are responsible for the distribution of federal grants and for corresponding management activities and advice to the BMBF within specified research areas.

Whereas federal research programs frequently deal with the promotion of science and technology development, state research programs more often refer to the implementation of technological capabilities within regional settings and thus usually do not interfere with federal programs.

With increasing globalisation of the economy and technology development in particular STP experiences contrasting demands to intensify scientific and technological development, on the one hand, and to minimise the risks of these new technologies, on the other hand. As a consequence, STP has deployed more levels of governmental action, depends on the co-operation with social actors outside the formal polity, and has developed new forms of policy impact assessment (technology assessment) and conflict management (mediation). Overall, a growing emphasis on (commercial) application and innovation as well as on networking among concerned social actors can be observed in STP, since the 1980s at least. The relative neglect of (scientific) education by STP has raised increasing concern in recent years, where the federalist German system of organizing science and education may well have contributed to this. In spite of this programmatic emphasis, the (federal) budget for research and technology experienced little growth since the 1990s, in contrast to considerable increase in the 1960s and 1970s, and the overall funding of R&D (above €40 billion after 1995) as percentage of GNP rose till 2,9% in Germany 1987-89, but declined to 2,3% after 1995 (BMBF, 1998).

4.3.2   Environment-oriented technology policy

Environment-oriented STP developed in parallel with EP since about the early 1970s (cf. Küppers et al., 1978; BMBW, 1972). From the beginning, environmental research oriented to supporting legal standards and provisions was mainly funded under the responsibility of the Ministry of the Interior[9] and of the formally subordinated UBA (Umweltbundesamt, federal environmental agency) established in 1974. Later this was handed over to the Ministry of Environment (BMU, established in 1986), whereas in general environmental research and technology development was mainly funded and organised by the BMFT/BMBF. Nevertheless, the (unavoidable) overlap in research objectives led to more or less continuous competition and ministerial departmental egotisms between these two ministries.

Furthermore, environmental research and technology development are funded and organised by state ministries (Environment, or Science and Technology), too, because the states have considerable jurisdiction over many environmental affairs. In addition, environmental research is also increasingly funded and commissioned by the research organisations and foundations, such as DFG, MPG, FhG and others, as well as by the European Commission. Since 1990 the DBU (Deutsche Bundesstiftung Umwelt;

---

[9]   The Ministry of Interior had the main responsibility for environmental affairs until 1986.

German environmental foundation) sponsors practice-oriented environmental research and technology development. In this context it is worth noting that the large (former nuclear) research centres had to be partly forced by quite some social and economic pressure to finally reorient their research capacities towards enhanced environmental research around 1990 in order to politically justify their further survival.

The financial figures of public expenditure for environmental research and technology development show an enormous increase over the last decades; from ca. €50 million around 1970, over €250 million around 1980, and €500 million around 1990, to €750 million in 1997[10], compared to about €33 billion overall expenditures for environmental protection measures in 1997. These figures represent only about 3-5% of all public R&D, and are relatively high in comparison with other EU countries.

In substance, the general objectives of ETP changed over time. In the beginning, the BMFT was only in charge for clean technologies and products, and for radiological protection within the federal environmental program of 1971, and mainly funded environment-oriented technology development projects according to industry-related priorities of media-specific environmental protection (Küppers et al., 1978). The (natural and engineering science) technology orientation of environmental research and technology programs of the BMFT/BMBF largely remained as dominant ETP objective helping to determine and improve the state of art of (science and) technology. But the funding of more basic environmental research (e.g. ecosystems research, climate research, research on species and nature protection) has increased considerably. A broader and more holistic perspective has characterised its conceptualisation and contextualisation in subsequent programs, though this has happened more on a symbolic than on a substantive level. Characteristic features of these programs are a stronger emphasis on the systematic, interdisciplinary and intermediary research of ecological systems and structures, coordinated development of respective environmental technologies, reduction and prevention of environmental burden, and readjustment of environmental damages. Production-integrated environmental protection (PIUS) fosters innovation towards new, environmentally benign paths of technology and product development, and the increase of ecological efficiency and productivity. On the agenda are also recently sustainable development of economic activities, and networking of

---

[10] The federal government contributed about €500 million, and the BMFT/BMBF share grew to about €350 million in 1997. In contrast, the BMU/UBA disposed of only about €28 million of research funds, indicating its relative junior position in environmental research and technology development, nowadays, as compared to its much more significant position in the 1970s.

STP programs and concepts (see BMFT, 1984, 1989; BMBF, 1997; Katz et al., 1997, 1998).[11]

So ETP primarily aimed at demonstration projects and plants advancing the state of the art of (environment-oriented) technology, largely in parallel to the development and passing of environmental laws by EP in the 1970s and 1980s. On the one hand, ETP strives much more for the development of general environment-oriented technologies aspiring to ecological optimisation in accordance with preventative EP goals, too, by supporting and influencing corresponding environmental research and management efforts in industrial corporation, where the deployment of environment-oriented technologies mainly takes place anyway. In particular, projects in the grey area receive funding where advantages and risks tend to outbalance each other. On the other hand, however, one should remain aware of the fact that genuine impulses for new environment-oriented concepts and technologies typically stem from non-governmental actors of the eco-scene and not from the BMBF or BMU, as could well be expected. If such new ideas do not fit the prevalent (technical-fix) orientation of the BMBF they tend to have a hard stand to be taken up and supported by the ministry and thus by formal ETP.

4.4   Co-ordination of EP and ETP in Germany

4.4.1   Actors and structural arrangements in EP/ETP co-ordination

The actors[12] who can be reasonably assumed to be possibly involved in EP/ETP co-ordination are those organised institutions, which are concerned with EP or STP. These may well be actors outside the formal polity, such as research institutes.

The actor constellation is as follows: as one would expect, the main actors are the BMBF and BMU, in particular those units charged with the task of co-ordination, and the UBA as institution subordinate to the BMU. All other actors are either much less engaged, mainly via the medium of working or advisory groups or only in a case-specific manner, or are hardly involved at all, though these may well undertake some internal co-ordination

---

[11]   Such an EP/ETP orientation was already recommended by the Picht-Commssion advising the government on the first federal environmental program, but not adopted by the federal government in 1971.

[12]   Political actors, i.e. social actors involved in or influencing politics, are typically organisations, but individuals, groups, and eventually meta-organisations, e.g. the state or industry, may be actors, too, although the latter ones usually should not be viewed as actors.

with respect to environmental research and environmental regulations. The first group of actors, partly engaged in general EP/ETP co-ordination, include those research project management bodies charged by the BMBF to manage funding and supervising environmental research projects. That is DLR, FZJ, FZK, GSF, and UBA, the UMK and its corresponding working groups of the federal and state governments, occasionally the cabinet, the parliamentary enquete commission 'Protection of man and the environment', and the TAB. In addition, research institutes and industrial corporations, involved in specific projects addressing both, environmental research and environmental regulations, and the VDI, involved in setting technical standards are also included (cf. Brennecke, 1996; VDI, 1991; Wolf, 1988). Other federal and state ministries that also fund environmental research and the umbrella research organisations and foundations, such as DBU, DFG, MPG, and FhG, belong to the second group of actors, those hardly involved in EP/ETP co-ordination. These actors more or less autonomously decide on the size and direction of environmental research and technology (with indirect relevance for environmental standards and regulations), which they fund.[13] They normally don't want to be influenced by the BMBF or BMU, and don't intervene in their affairs, either.

Overall, there is some gradual increase in EP/ETP co-ordinating activities over the past two decades, essentially between BMFT/BMBF and BMU. These remain largely on a formal level (obligatory early departmental co-ordination since 1975 for projects above €100.000) and on occasion. Both ministries take well notice of each others activities, and they partly compete on the lead role in environmental strategy and technology development. However, even the individual chapters of their first joint environmental research program, initiated by common declaration of both ministers and finally agreed upon after two years of debate in 1997, are largely written by the one or the other ministry.[14]

Basically, the BMBF is aware of its formally limited role in EP and accordingly tries to exert indirect influence in this respect by propagating technology options favourable for certain environmental regulations and by mediating networks of environmental research programs and technology development. So EP/ETP co-ordination mainly concerns the design, implementation and regulatory impact of ETP programs. As a consequence it

---

[13] DBU-funded projects are notified to the BMBF and BMU which may contradict a project, at least in principle.

[14] Already in 1971 the Picht-Commission saw lacking co-ordination between (federal) ministries as the main deficit to install a systematic and comprehensive environmental (research) program and policy. At that time, departmental structures were dominant in formulating the federal environmental program, too, so that a comprehensive concept of environmental protection or even utilisation was bound to fail (Küppers et al., 1978: 147ff).

is important to distinguish between the formal policy and the technical expert level of EP/ETP co-ordination. To some degree at least, underlying power games among the actors involved about their relative position and a favourable pattern of interest consideration tend to play a much more prominent role on the first than on the second level. Technical experts also tend to bring in more substantive new ecological ideas and concepts and to be more fact-oriented than policy makers.

So, apart from (case-specific) co-operation of technical experts (in corresponding working groups), the dominant mode of actor co-ordination is one of notification, of demarcation to protect and keep one's own (policy) domain, of participation in advisory groups, of eventual mutual co-ordination of activities, and of initiating complementary measures by other actors. Consequently, one may speak of partial and occasional EP/ETP consultation and co-ordination, but hardly of inter-policy co-operation.

This can be illustrated by few typical actor configurations:

1. With the increasing number of research and technology projects the continuous information of other ministries concerned about pending project approvals usually will not lead to mutual consultation due to the already quite limited capacity of responsible units within a ministry.

2. Compared with other ministries the BMBF sees itself, on the one hand, in the privileged position of being able to take up or to initiate whatever topic it considers relevant for research or technology promotion. Therefore it takes a more long-term and strategic perspective implying a lead role in ETP, and to act - without the threat of legal intervention - as a mediator in networks of environmental research programs and technology development, where industrial corporations, scientific institutes, technical associations and regulatory bodies participate. It considers these consultation and co-ordination processes of launching and implementing environmental research and technology programs as well working ones with rather fair distribution of tasks and influence among the participating actors. Although the BMBF views the results of its research programs and projects, e.g. concerning environmental indicators or dying forests, as serious impulses for EP, it is, on the other hand, not much interested and involved in actual EP, and therefore was hardly engaged for instance in the development of the environmental statute-book. So ETP impacts on EP largely remain indirect ones.

3. In this context the BMBF also tries to achieve substantial involvement of other ministries concerned by making them responsible for organising co-ordinating working groups and thus generating corresponding identification effects, e.g. having the Ministry of Transport (Bundesverkehrsministerium) in charge of co-ordinating the environmental research program on the Elbe river.

4. The BMU, and even more so the UBA with diminishing research funds see themselves in a relatively dispriviledged position in ETP, being insufficiently consulted in the definition and bargaining processes of environmental research programs and technology development co-ordinated by the BMBF. Thus they do not really see themselves considered as co-operative partners in program related advisory bodies.

5. The various project management bodies know the relevant research landscape and therefore are usually well able to manage the review and selection procedures of corresponding environmental research programs. So they have quite some influence on the concrete direction of environmental research although the BMBF occasionally may well decide differently about some proposals. Although these bodies know about corresponding EP issues and problems, they are usually not engaged in EP/ETP co-ordination efforts. An exception is the TÜV Rheinland (earthbound transportation technologies), the UBA (waste management and cleaning-up of waste residues), or the VDI (physical technologies, laser research and technology). Even in these cases, however, no systematic, but only indirect informal and coincidental co-ordination efforts can be observed, since both tasks are usually considered as relatively independent on the level of concrete action and decision-making, and frequently handled by different units of the organisation.

6. On the level of federal states ETP, comprising ca. €250 million in 1997, is (nowadays) mostly pursued in a rather issue- and case-specific way oriented towards improvements of existing installations and production processes. So the environmental ministries together with responsible environmental (state) agencies typically fund corresponding development activities in exemplary companies in order to achieve certain environmental protection measures or standards agreed upon with an industrial sector or already prescribed, which have to be adopted by all its companies afterwards. Thus, at least in part ETP appears to be relatively well co-ordinated with and organisationally integrated in EP on the state level, after principal environmental legislation and technologies have been issued and developed in the 1970s and 1980s. Overall, EP/ETP co-ordination seems to be developed stronger in environmental policy fields oriented towards end-of-pipe solutions, which are typically combined with a preference for command-and-control approaches and fall mainly under the responsibility of state governments. Furthermore, considerable emphasis is put on cooperative agreement and action of government programs with industry. However, little consultation and co-ordination of state ETP appears to take place with federal ETP, as they proceed

relatively independently of each other and follow different objectives in every-day practice.[15]

7. The large umbrella research organisations - DFG, MPG, FhG and DBU - each spending meanwhile more than €50 million annually on environment-related research projects, largely decide on their own and don't interfere with direct federal environmental research programs.

4.4.2   Evolution and levels of EP/ETP co-ordination

The evolution of relevant boundary conditions of EP/ETP co-ordination may be summed up by the following key development trends. Concerning ETP, its gradual development over the last decades can be observed on the cognitive, social, economic, and institutional level with changing points of reference: technical (end-of-pipe) solutions, integrated (environment-oriented) technologies, sustainable development, innovation and globalisation. Similarly, EP follows a more preventive and integrated orientation (symbolically) referring to the principles of prevention, polluter-pays and co-operation as well as to sustainable development (BMU, 1998). This has been possible because the legal, institutional, financial, and technological basis to deal with severe and acute environmental problems more effectively has been established to a considerable degree; but also because serious substantive environmental management efforts have been developed in companies as well as in other organisations since around 1990. STP, and within it ETP, are mostly separately formulated and pursued from EP. Therefore, it will be quite interesting to trace (partial) EP/ETP policy co-operation, let alone integration, in (innovation) networks within well specified case studies.

With respect to different levels of inter-policy co-operation (Conrad, 2000a), one finds non-co-operation, mutual regard of relevant policies, and to some degree conscious co-ordination of policy programs and of specific projects between EP and ETP. The major form of EP/ETP co-ordination was, and still mostly is an indirect (and informal) one. In addition, however, the selective inclusion of BMU/UBA officials in working and advisory groups of ETP and the already mentioned process of formalised obligatory early departmental co-ordination for projects above €100.000 have to be mentioned, which implies (early) information, consultation on request, and bargaining and co-ordination in case of opposition. Due to mutual regard of relevant policies, probable requirements and impacts of EP for ETP, or of ETP for EP are well recognised and (informally) communicated. As a result hardly any systematic but mainly situational case-specific co-ordination

---

[15] See Kern 1997 on the subject of policy innovation in multi-level systems.

(efforts) can be observed, accentuating policy co-ordination on the level of program implementation.

In former decades the needs of EP, for instance setting emission or ambient air or water quality standards, induced the promotion of corresponding research and technology development activities in ETP, in accordance with the predominant favour for corrective technical solutions of environmental hazards. ETP had mainly the task of contributing to improving the state of the art of (science and) technology formulated in the respective norm and standard setting committees. Similarly, according to the availability (and economic viability) of environment-oriented technologies, corresponding environmental regulations were formulated and passed with stronger or weaker rules, frequently demanding gradual adoption of stricter standards. This did not imply an instrumentalisation of ETP by EP; ETP was only aware of EP needs and therefore responded to them or took itself initiative by promoting favourable technological development projects, which had impacts on EP regulations (see ECOTEC, 1988; Coenen, 1991; Angerer et al., 1996, 1997; UBA, 1998).

Meanwhile, with changing perception, conceptualisation, and handling of environmental problems the options and development paths of EP and ETP are more flexible, manifold and complex so that each policy may induce new developments which need co-operative efforts of the other, but without clear specification of these requirements. For instance, stimulating environment-oriented innovations or sustainable economy projects or climate research has no clear-cut impacts on EP regulations. And voluntary agreements and obligations of industry to fulfil certain EP goals, e.g. reducing environmental pollution or installing closed production cycles, do not yet imply specific ETP requirements. They still have to be defined in the regulatory network (cf. Van Dijken et al., 1999). Thus, a more holistic ecological perspective of EP and an increased funding of environment-related research and technology development formally require enhanced inter-policy co-operation in order to effectively take up and realise (new) environment-oriented ideas and solutions typically originating from scientific and technical experts outside the genuine sphere of politics. In spite of rather unavoidable accompanying power games of political actors one can observe more co-ordination efforts as opposed to explicit conflict performance. Apart from few counter examples enhanced EP/ETP co-ordination usually does not involve (potentially well feasible) trade-offs with effective environmental protection.

So EP and ETP presently both influence each other on a rather equal level due to substantive reasons and due to the now prominent role of the BMBF in ETP. EP/ETP co-operation has increased on the level of mutual regard of relevant policies and of conscious co-ordination of specific projects. In spite of the possibility of further improvement and of favourable win-win situations one should be aware of the differing objectives and of differing policy styles, profiles and instruments of EP and ETP,[16] where ETP is typically more interested in pushing environment-oriented innovations and EP more in supporting their diffusion. Thus, the scope of, the interests and influence in EP/ETP co-ordination can be expected to remain limited ones. As a tendency, they may increase due to the differentiation and complexity, which the social, political and economic treatment of environmental hazards has reached. Secondly, co-ordination may occur due to common worldviews of the actors involved, for instance production-integrated environmental protection, holistic consideration of environmental problems, and sustainable development as guiding principles of EP and ETP. Finally due to case-specific co-ordination activities, leaving aside the probably irreducible (institutional and personal) endeavour to influence and determine other policies along the lines of one's own concepts and interests.[17]

Finally, three case-specific examples are briefly described in this section, in which EP/ETP co-ordination took largely place on the level of implementing policy programs and led to successful environmental improvements with the help of environment-oriented technology development and environmental standards and regulation. These exemplary cases are the phasing out of CFCs, the phasing out of diluted acid dumping, and the substitution of chlorine bleaching by oxygen and peroxide bleaching in the pulp and paper industry. This description gives no analysis of the process of EP/ETP co-ordination, but provides at least some plausibility for the above mentioned prevailing indirect mode of co-ordination, based on a common understanding of environmental problem solutions.

In view of the diminution of the stratospheric ozone layer by CFC emissions the ozone regime as one of the few effectively functioning international environmental regimes organised the general phasing out of CFCs (Vienna Convention, 1985; Montreal Protocol, 1987, etc). German EP followed suit and in 1991 passed the CFC/Halon Prohibition Ordinance (as part of the chemical act), regulating transition periods and forbidding the use of CFCs after 1995 at the latest, and the FC Ordinance (2. BImSchV, 2. Verordnung zum Bundes-Immissionsschützgesetz), regulating the diminished utilisation of FC solvents in surface cleansing, cleaning of textiles, and extraction installations. The parallel ETP activities strongly helped that industry could keep to the prescribed periods by promoting the development of various CFC substitutions or related technologies, especially

---

[16] For instance, if an environment-oriented technology (promoted by ETP) does not prove to be economically viable then EP has to find a (regulatory) solution of sufficient environmental protection, but this is not considered the task of the BMBF.

[17] This is illustrated by the already mentioned formulation of the first common environmental research program of BMBF and BMU.

since 1987/88. In 1991 altogether 52 corresponding projects were funded, 35 of them aiming at CFC substitution. 34 projects contributed to the arrangement and implementation of both ordinances, with 15 projects making essential or important contributions (Angerer et al., 1996, 1997).

In 1977 the Hohe See Einbringungsgesetz (Open Water Dumping Act) limited dumping of wastes in the sea to those cases which are either ecologically acceptable or not possible on land. Diluted acid originates as hazardous waste from the sulphate process producing titanium-oxide and was dumped in the North Sea. Technological developments promoted by the BMFT helped to establish a closed cycle management of diluted acid and a land-based utilisation of organically polluted waste sulphuric acid. Therefore, dumping of diluted acid was stopped in 1989 and the necessary corresponding legal permission is no longer given (ECOTEC, 1988; Angerer et al., 1996).

In the 1970s and 1980s the BMFT funded a considerable number of environment-oriented technology projects in the pulp and paper industry, though with limited success in the end. Furthermore, the UBA co-funded demonstration plants in the 1980s to demonstrate the technical feasibility to drastically reduce emission of organic chlorine compounds (AOX) by oxygen and peroxide bleaching in pulp mills. At first the German pulp and paper industry strongly opposed effective environmental regulations for more than a decade, but a waste water levy on effluents has induced considerable efforts of emission reduction. When a new AOX-standard for waste water in the pulp and paper industry became effective in 1990,[18] the industry mostly changed to totally chlorine-free (TCF) bleaching or closed some pulp mills within a year (ECOTEC, 1988; Angerer et al., 1996, 1997; Conrad, 2000b).

4.4.3   Determinants of EP/ETP co-operation

What is the pattern of determinants of the development of EP/ETP co-operation? From the analytical perspectives it appears reasonable to distinguish the following types of determinants. *One key factor* is the general (perceived) *problem structure* which EP and ETP each address: how similar are the (environmental) problems to be tackled; can they be easily linked; how complex and differentiated are both policy arenas? Thus, for instance, the common orientation towards technological solutions to environmental problems and the obvious link between available technical options and environmental standards was certainly favourable for EP/ETP co-ordination in the past. *A second key factor* is the (implicit) *cost-benefit calculation* of the actors facing EP/ETP co-operation: what are the probable substantive

---

[18]   Abwasserverwaltungsvorschrift (the 19th waste water administrative ordinance, part A).

and political benefits of co-operation; what are the likely opportunities costs; how viable is the co-operation? If for instance favourable contacts and dialogue conditions with industrial sectors and corporations by the BMBF are threatened by the BMU because of envisaged new environmental regulations, it may already be doubtful for the BMBF to seek close co-operation with the BMU in this case. In the political system in particular, the cost-benefit calculation especially depends on the estimated gains or losses of power by investing in inter-policy co-operation, on the one hand, and on the subjective self-images and the images of eventually collaborating organisations or persons. So interestingly both, the BMBF and the BMU/UBA, tend to see themselves as environmental front-runners and the other one as lame duck in this respect, what - in accordance with the already mentioned Thomas theorem - clearly has an impact at least on the atmosphere of EP/ETP co-operation.

Certainly the degree of (structurally entrenched) *departmental egotism* decisively influences success or failure of EP/ETP co-operation.[19] Here the situation is determined by both, the actual degree of departmental egotism and self-interest and, again, its subjective perception by the other actors, as indicated in section 4.4.1 for BMBF and BMU/UBA. Furthermore, the (institutional and situational) *political context* strongly influences the viability of EP/ETP co-operation. How much is inter-policy co-operation supported by the government in general and in specified circumstances; how much are other ministries concerned about and involved in EP/ETP co-ordination, e.g. transportation, agricultural, and economic policy. What kind of (departmental) advocacy coalitions shape the actual degree of EP/ETP co-operation? The BMBF may well form an alliance with the Ministry of Economic Affairs to foster its deviating position against the BMU, as it occasionally happened in the past. In general, *contextual determinants* are highly important for (successful) EP/ETP co-operation: the role of key figures and their world view and interests, the availability of resources for co-ordination activities[20], the degree of personnel fluctuation and exchange, the general political atmosphere and policy style with respect to ecological questions, the socio-cultural climate, particularly environmental awareness of the population, and the development of public opinion and debate. How much these individual factors actually influence specific cases of EP/ETP co-operation is an empirically open question. As confirmed by several

---

[19] This structural problem might change but would not vanish if EP and ETP responsibilities would be put together in one ministry because similar questions of competence and competition would arise between the different units of such a ministry.

[20] Scare (public) resources will tend to undermine the feasibility of inter-policy co-operation and the time needed to allow for reflexion and substantive exchange and communication necessary for its effectiveness.

studies and by some interviewees, the social significance of environmental questions clearly influences the political strength of EP actors and therefore the pressure to achieve viable EP/ETP co-operation. Similarly, a cultural climate in favour of environmental protection provides incentives for scientists and technology developers to generate (new) scientific and technical solutions to cope with environmental problems. Finally, globalisation of the economy renders technically rather simple environmental improvements, such as the elimination of stand-by losses in electrical appliances or the passing of the electronic waste ordinance (Elektronikschrott-Verordnung), difficult to realise on the socio-political level. This is so because such devices are produced and imported from all over the world and the necessary concerted political action is hardly feasible due to diverging perceptions and interests of policy makers from different countries, or even only ministries.

A particular role may be played by *historiographic particularities* (in the sense of historically and situationally contingent chance qualifiers). Whereas the general decree of obligatory early departmental co-ordination for projects above €100.000, issued in 1975, clearly mattered, the change in government in 1982 seemed to have had only little influence on EP/ETP co-ordination efforts. Significant influence has to be attributed to the *self-dynamics* and corresponding *trickle down effects* of inter-policy co-operation. Self-dynamics refer to past experiences with other actors, and the experienced willingness to co-operate by part of the personnel. The dynamics include also learning effects, push-and-pull effects of incentives set by the minister or leading persons, institutionalised co-ordination routines, favourable incentives set by win-win constellations, organisational learning, and time spent for reflection on and training in organising and implementing co-operation. Trickle down effects relate to the development of (successful) policy co-ordination: from common problem perception to mutual information to partly common working groups to joint policy programs to mutual project co-ordination to partial inter-policy co-operation, and to actual elements of policy integration.

Again, it remains an empirically open question as to how far and how rapidly such a dissemination process actually develops. For EP/ETP co-ordination it appears to remain a relatively limited one.

These seven types of determinants of EP/ETP co-operation do not denominate specific substantial determinants but rather mainly structural types of possible influencing factors, considered analytically reasonable. The plausibility of this list invalidates its arbitrariness but does not justify it in theoretical terms. To determine the relative weight of individual substantive factors of influence on EP/ETP co-operation in Germany goes beyond the scope of this chapter.

In the end, the interplay of the various determinants decides about the actual form and development of EP/ETP co-operation. How this interaction dynamics does look like usually is a case-specific phenomenon and can therefore hardly be generalised in theoretical terms.

4.4.4   Fundamental and manageable problems of EP/ETP co-operation

When analysing or demanding for better EP/ETP co-operation, one should distinguish between manageable and fundamental effects and problems of inter-policy co-operation in order not to aspire at impossible objectives. So even under the idealistic assumptions that all relevant actors are willing to collaborate and that satisfactory institutional arrangements to cooperate are available, some negative effects of inter-policy co-operation cannot be overcome on principle.

On the one hand, beyond optimal level additional benefits of further enhanced policy co-ordination/co-operation/integration diminish and are outbalanced by its growing costs of time and manpower resources. On the other hand, given the always limited capacity of any (formal) organisation, over-proportional emphasis on inter-policy co-operation in the longer run necessarily implies qualitatively the watering down of its genuine duties and quantitatively the neglect of its other relevant tasks. There was and is always demand for better communication and co-operation between neighbouring policy fields, between science and politics, or between science and industry; and this demand is always bound to limited success and failure. So this situation is not specific for EP and ETP.

Furthermore, in the real world of politics interests in power gains and subjectively biased (self-) images of actors preclude such ideal-type settings. Actor strategies and policy games underlying attempts of inter-policy co-operation therefore may well be reconstructed in terms of game theory, as done for instance by Brennecke (1996) in his analysis of the interplay of governmental steering and societal regulation concerning standard setting.

However, within such basic limitations there is still considerable scope for improvements in EP/ETP co-ordination. Besides possibilities to induce a political atmosphere and a policy style favourable towards inter-policy co-operation, and to reduce the prominence of departmental egotism, some problems are manageable. Thus, problems of scarce resources and time constraints for cooperative activities can well be surmounted which hinder individuals and units to become motivated and to get some training in this respect, to undergo learning processes, and to develop creative ideas and a more reflexive stance. Limiting personnel fluctuation, on the one hand, and allowing for (temporary) personnel exchange between co-operating organisations, on the other hand, will also tend to favour EP/ETP co-ordination. Finally, some self-supporting dynamics of EP/ETP co-operation can be induced by positive (past) experiences and by the motivating effects of key individuals who are involved. Speaking of manageable problems of EP/ETP co-ordination assumes that they can be seriously diminished in optimisation processes indicated above, but not completely removed.

4.5   EP, ETP and environment-oriented innovation: some conclusions

The reason of this chapter to describe and to analyse prominent features, development and problems of EP/ETP co-ordination in Germany is to provide background knowledge for the investigation of EP and ETP influence in networks of environment-oriented innovation.

Innovations are typically differentiated in technical (process or product) innovations and organisational innovations, and in basic and incremental innovations. Although it is possible to denominate success factors of innovation,[21] innovation is per definition a project under conditions of uncertainty with uncertain outcome (Dodgson and Rothwell, 1994; Dosi et al., 1988; Freeman, 1992; Lundvall, 1992; Meyer-Krahmer, 1998; Nelson, 1993).

> *"Success is a matter of competence in all functions and of balance and co-ordination between them. Finally, success is, 'people centred' and, while formal techniques can enhance the performance of dynamic, gifted and entrepreneurial managers, they can do little to raise the performance of innovatory management lacking these qualities." (Rothwell, 1994: 37)*

---

[21] Rothwell (1994: 36) provides the following list of project execution and corporate level factors: Good internal and external communication: accessing external know-how. Treating innovation as a corporate-wide task: effective inter-functional co-ordination: good balance of functions. Implementing careful planning and project control procedures: high quality up-front analysis. Efficiency in development work and high quality production. Strong marketing orientation: emphasis on satisfying user needs: development emphasis on creating user value. Providing a good technical service to customers: effective user education. Effective product champions and technological gatekeepers. High quality, open minded management: commitment to the development of human capital. Attaining cross-project synergies and inter-project learning; Top management commitment and visible support for innovation. Long-term corporate strategy with associated technology strategy. Long-term commitment to major projects (patient money). Corporate flexibility and responsiveness to change. Top management acceptance of risk. Innovation-accepting, entrepreneurship-accommodating culture.

Although external information networks and collaboration with users was always of vital importance for the development of new products and processes, systems integration and networking now tend to play a crucial role in national and international innovation systems. These encompass greater overall organisational and systems integration, flatter, more flexible organisational structures for rapid and effective decision-making, fully developed internal data bases, and effective external data links as their primary enabling features (Freeman, 1991; Rothwell, 1994).

Environment-oriented innovations aim at the prevention and diminution of environmental degradation by anthropogenic activities, the repair and removal of environmental damages, and the diagnosis and control of environmental burden. Environment-oriented process-, product- or organisational innovations influence the level of integrated, end-of-pipe, and organisational environmental protection. Environmental protection is only one out of many innovation goals of a company within its overall innovation processes, within which the weight of environmental protection goals and their complementarity or competition with other innovation goals determine its role and importance. The more integrated environmental protection measures are, the less can environment-oriented innovations be isolated from 'normal' innovations.

Compared with normal innovations, generating environment-oriented innovations strongly depends on the existence and appropriate arrangement of environmental policy (instruments), because market signals frequently don't correspond to the socio-economically wanted importance of environment-oriented innovations. Therefore decisions on environment-oriented innovations are characterised by even more uncertainty and ignorance than 'normal' innovations. In addition, the impacts of environmental regulation on innovation are determined by the interaction of EP with other determinants of the innovation process. They thus cannot be determined in generalised terms (cf. Hemmelskamp, 1997; Coenen et al., 1996; Fischer and Schot, 1993; Gleich et al., 1997; Kemp, 1997, 1998; Klemmer et al., 1999; Minsch et al., 1996; Wallace, 1995). Since environment-oriented innovations have in principle the potential both to solve (known) environmental hazards and to create new (unknown) ones (see the example of genetic engineering), and since a more comprehensive holistic perspective of environmental problems within the frame concept of sustainable development does not deal with them in isolation, EP should not focus too narrowly on environment-oriented technology and innovation with its inherent end-of-pipe orientation. Rather it should focus on outcomes, not on specific technologies and support the development of technological change in general into an ecologically benign direction (Banks and Heaton, 1995; Porter and Van der Linde, 1995).

As environmental concern, research, policy, legislation, and management evolved over phases with differing degrees of comprehensiveness, internal differentiation, social and institutional stabilisation, and of capacity to achieve their objectives, effective EP/ETP co-operation cannot be expected to already possess a satisfying socio-structural basis and to develop rapidly. Instead, one may be better off relying on the decade-long trickle down effects in the development of (successful) policy co-ordination mentioned above.

What should one then expect - in view of general theoretical and empirical knowledge about innovation, economics, politics, and policy in modern societies - concerning the role of EP, ETP, and EP/ETP co-operation in corporate environment-oriented innovations?

1. Corporate innovations typically happen without substantive interference of politics. Some company external actors are frequently involved in innovation networks, but rarely actors of the political system. Policy may well provide favourable or unfavourable (regulatory) boundary conditions, but usually will not contribute in substance to specific innovation processes.[22]

2. Although EP and partly TP typically aim first at developing and defining the regulatory framework shaping processes of environment-oriented innovation in general, they may push (or restrict) specific environment-oriented innovations, too. Such policies have to be more or less tailor-made in order to be successful (Van Dijken et al., 1999) and require considerable financial and personnel resources (see the example to abolish dumping of diluted acid in the North Sea). They will remain rather the exception than the rule, already because of the limited capacity of the political system to substantively control and steer (economic industrial) innovation.

3. For similar reasons, inter-policy co-operation usually will remain rather the exception than the rule, too. Therefore, even if EP or ETP are engaged substantively in a specific innovation process, EP/ETP co-operation will not happen without conscious attempts of political actors in both policy fields in this direction.

4. Thus, EP/ETP co-operation in particular will on average have only an indirect impact on (success or failure of) environment-oriented innovation. That is not to deny the real possibility of significant EP/ETP co-operation in environment-oriented innovation networks, but to point out their relative improbability for simple (socio)structural reasons.

---

[22] Typically, (E)TP is interested to induce the generation of (environmental) innovations, and EP in their diffusion.

When one analyses and summarises (typical) empirical case studies of environment-oriented innovations (cf. Klemmer, 1999; Blazejczak et al., 1999; Conrad, 1998; Jacob, 1999; Jacob and Jänicke, 1998; Victor et al., 1998), the findings tend to confirm the above expectations. For the majority of cases EP or ETP either play no role or have more or less influence by providing significant regulatory or financial boundary conditions. For a considerable number of cases EP or ETP have been substantially involved in the innovation process, too, but rarely systematic and not only incidental EP/ETP co-operation could be observed.

Nevertheless, its role should not be underestimated. In practice, with the gradual installation of environmental management schemes in companies (and public institutions), positive effects of EP and ETP (co-ordination) are possible and will depend on the establishment of appropriate arrangements and routines indicated above. On this level, the comparative perspective may well provide insights about the relative status and feasible improvements in EP/ETP co-ordination and their role in environment-oriented innovation.

So inter-policy co-operation (of EP and ETP) is feasible in a productive manner, at least this is so in principle. However, one should be strongly aware of the limitations of inter-policy co-operation, as pointed out, and not overburden this approach. In practice, one will always find an overlap of all forms of (positive and negative) interaction between two policy arenas.

Given different (socioeconomic) interests with differing power as well as certain institutional interests in policy-making (according to the type of game, decision rules and decision styles; Scharpf 1985, 1989, 1994), inter-policy co-operation should enable to experiment with various models of co-operation which foster learning capabilities of the actors willing to co-operate.

Under the present conditions of social change on various levels, socio-cultural orientations tend to become crucial determinants for the success of a social project. Thus, the emphasis on social learning processes in inter-policy co-operation appears reasonable, which justifies the prominence of moral persuasion over strong regulatory policy instruments. With the loss of commitment of formal policy rules and routines, confidence (building), anchored in personal relations, gains central importance for the viability and reliability of inter-policy co-operation. More generally speaking, socio-psychological factors play a more important role again in politics under conditions of continuously changing environments involving a growing fluidity of interests and motivations of individuals and corresponding threats to the sovereignty and identity of formal organisations.

As a consequence, this implies that successful inter-policy co-operation primarily depends on a lot of case-specific practical work by the actors involved. This is because of the specifity and complexity of substantive environment-related projects, on the one hand, and because of the significance of intentional individual action for an effective EP/ETP cooperation, on the other hand.

REFERENCES

Angerer, G. et al. (1996) *Einflüsse der Forschungsförderung auf Gesetzgebung und Normenbildung im Umweltschutz.* Heidelberg: Physica.

Angerer, G. et al. (1997) *Wirkungen der Förderung von Umwelttechnologie durch das BMBF.* Bonn: Report.

Babel, C., Zschörnig, B. (1999) Abschätzung der innovativen Wirkungen des Umwelthaftungsrechts - dargestellt am Beispiel des Umwelthaftungsgesetzes. In: P. Klemmer (ed.) *Innovationen und Umwelt. Fallstudien zum Anpassungsverhalten in Wirtschaft und Gesellschaft.* Berlin: Analytica.

Banks, R.D., Heaton, G.R. (1995) An innovation-driven environmental policy. *Issues in Science and Technology,* 12 (1): 43-51.

Blazejczak, J. et al. (1999) Umweltpolitik und Innovation: Politikmuster und Innovationswirkungen im internationalen Vergleich. In: P. Klemmer (ed.) *Innovationen und Umwelt. Fallstudien zum Anpassungsverhalten in Wirtschaft und Gesellschaft.* Berlin: Analytica.

BMBF (1996) *Bundesbericht Forschung 1996.* Bonn.

BMBF (1997) *Forschung für die Umwelt.* Bonn.

BMBF (1998) *Faktenbericht 1998 zum Bundesbericht Forschung.* Bonn.

BMBW (1972) *Forschungsbericht IV der Bundesregierung.* Bonn.

BMFT (1984) *Umweltforschung und Umwelttechnologie. Programm 1984-1987.* Bonn.

BMFT (1988) *Bundesbericht Forschung 1988.* Bonn.

BMFT (1989) *Umweltforschung und Umwelttechnologie. Programm 1989 bis 1994.* Bonn.

BMU (1990) *Umweltbericht 1990.* Bonn.

BMU (1994) *Umweltpolitik. Umwelt 1994. Politik für eine nachhaltige, umweltgerechte Entwicklung.* Bonn.

BMU (1998) *Umweltpolitik. Umweltbericht 1998.* Bonn.

Braun, D. (1997) *Die politische Steuerung der Wissenschaft.* Frankfurt: Campus.

Brennecke, V.M. (1996) *Normsetzung durch private Verbände. Zur Verschränkung von staatlicher Steuerung und gesellschaftlicher Regulierung im Umweltschutz.* Düsseldorf: Werner-Verlag.

Coenen, R. (1991) *Die Reaktion der deutschen Forschungs- und Technologiepolitik auf die Umweltproblematik in den 80er Jahren.* KfK 4804, Karlsruhe.

Coenen, R. et al. (1996) Integrierte Umwelttechnik - Chancen erkennen und nutzen. *Studien des TAB,* Vol.1. Berlin: edition sigma.

Conrad, J. (ed.) (1998) *Environmental management in European companies. Success stories and evaluation.* Amsterdam: OPA.

Conrad, J. (2000a) Interpolicy Coordination in Germany: Environmental Policy and Technology Policy. *Zeitschrift für Umweltpolitik und Umweltrecht,* 23: 583-614.

Conrad, J. (2000b) *Ecologically sound pulp production: how the interaction of world market conditions, corporate capability and environmental policy determines success and failure of environmental innovations.* FFU report 00-05. Berlin: Forschungsstelle für Umweltpolitik, Freie Universität Berlin.

Dijken, K. van et al. (1999) *Adoption of environmental innovations.* Dordrecht: Kluwer.

Dodgson, M., Rothwell, R. (eds.) (1994) *The handbook of industrial innovation*. Aldershot: Edward Elgar.

Dosi, G. et al. (eds) (1988) *Technical change and economic theory*. London: Pinter.

ECOTEC (1988) *Umsetzung/Nutzen der BMFT-Förderung Umwelttechnik*. München: Report.

Enquete-Kommission "Schutz des Menschen und der Umwelt" (1994) *Schutz des Menschen und der Umwelt. Die Industriegesellschaft gestalten. Perspektiven für einen nachhaltigen Umgang mit Stoff- und Materialströmen*. Bonn: Economica.

Fischer, K., Schot, J. (eds.) (1993) *Environmental strategies for industry*. Washington D.C.: Island Press.

Freeman, C. (1991) Network of innovators: a synthesis of research issues. *Research Policy,* 20: 499-514.

Freeman, C. (1992) *The economics of hope*. London: Pinter.

Gleich, A. von et al. (eds) (1997) *Surfen auf der Modernisierungswelle? Ziele, Blockaden und Bedingungen ökologischer Innovation*. Marburg: Metropolis.

Hemmelskamp, J. (1997) Umweltpolitik und Innovation - Grundlegende Begriffe und Zusammenhänge. *Zeitschrift für Umweltpolitik & Umweltrecht,* 20: 481-511.

IIUW (1998) *Towards an integration of environmental and ecology-oriented technology policy. Stimulus and response in environment related innovation networks* (ENVINNO). Project proposal, Vienna: Interdisziplinäres Institut für Umwelt und Wirtschaft.

Jacob, K, (1999) *Innovationsorientierte Chemikalienpolitik. Politische, soziale und ökonomische Faktoren des verminderten Gebrauchs gefährlicher Stoffe*. München: Herber Utz.

Jacob, K., Jänicke, M. (1998) Ökologische Innovationen in der chemischen Industrie: Umweltentlastung ohne Staat? Eine Untersuchung und Kommentierung zu 182 Gefahrstoffen. *Zeitschrift für Umweltpolitik & Umweltrecht,* 21: 519-547.

Jänicke, M., Weidner, H. (1997) Germany. In: M. Jänicke, H. Weidner (eds.) *National Environmental Policies. A Comparative Study of Capacity-Building*. Berlin: Springer.

Jänicke, M., Kunig, Ph., Stitzel, M. (1999) *Lern- und Arbeitsbuch Umweltpolitik. Politik, Recht und Management des Umweltschutzes in Staat und Unternehmen*. Bonn: Dietz.

Jänicke, M. et al. (1992) *Umweltentlastung durch industriellen Strukturwandel?* Berlin: edition sigma.

Katz, Ch. et al. (1997) Monitoring "Forschungs- und Technologiepolitik für eine nachhaltige Entwicklung", *TAB-Arbeitsbericht,* 50. Bonn.

Katz, CH. et al. (1998) "Forschungs- und Technologiepolitik für eine nachhaltige Entwicklung", *TAB-Arbeitsbericht,* 58. Bonn.

Keck, O. (1993) The National System for Technical Innovation in Germany. In: R. Nelson (ed.) *National innovaton systems. A comparative analysis*. New York: Oxford University Press.

Kemp, R. (1997) *Environmental policy and technical change*. Aldershot: Edward Elgar.

Kemp, R. (1998) *Environmental regulation and innovation. Key issues and questions for research*. Ms. Maastricht.

Kern, K. (1997) *Die Diffusion von Politikinnovationen in Mehrebenensystemen. Politikintegration und -innovation in der US-amerikanischen Umweltpolitik*. Diss. Berlin.

Klemmer, P. (ed.) (1999) *Innovationen und Umwelt. Fallstudien zum Anpassungsverhalten in Wirtschaft und Gesellschaft*. Berlin: Analytica.

Klemmer, P. et al. (eds.) (1999) *Umweltinnovationen. Anreize und Hemmnisse*. Berlin: Analytica.

Küppers, G. et al. (1978) *Umweltforschung - die gesteuerte Wissenschaft?* Frankfurt: Suhrkamp.

Lundvall, B-A. (ed.) (1992) *National systems of innovation. Towards a theory of innovation and interactive learning*. London: Pinter.

Meyer-Krahmer, F. (ed.) (1998) *Innovation and sustainable development. Lessons for innovation policies*. Heidelberg: Physica.

Minsch, J. et al. (1996) *Mut zum ökologischen Umbau*. Basel: Birkhäuser.

Nelson, R.R. (ed.) (1993) *National innovaton systems. A comparative analysis*. New York: Oxford University Press.

Porter, M.E., Linde, C. van der (1995) Green and competitive. Ending the stalemate. *Harvard Business Review*, 9/10: 120-134.

Rothwell, R. (1994) Industrial Innovation: Success, Strategy, Trends. In: M. Dodgson, R. Rothwell (eds.) *The handbook of industrial innovation*. Aldershot: Edward Elgar.

Scharpf, F.W. (1985) Die Politikverflechtungs-Falle: Europäische Integration und deutscher Föderalismus im Vergleich, *Politische Vierteljahresschrift*, 26: 323-356.

Scharpf, F.W. (1989) Decision Rules, Decision Styles and Policy Choices, *Journal of Theoretical Politics*, 1: 149-176.

Scharpf, F.W. (1991) Die Handlungsfähigkeit des Staates am Ende des 20. Jahrhunderts, *Politische Vierteljahresschrift*, 32: 621-634.

Scharpf, F.W. (1994) Games real actors could play. Positive and negative co-ordination in embedded negotiations, *Journal of Theoretical Politics*, 6: 27-53.

Simonis, U.E. (ed.) (1988) *Präventive Umweltpolitik*. Frankfurt: Campus.

SRU (1994) *Umweltgutachten 1994. Für eine dauerhaft-umweltgerechte Entwicklung*. Stuttgart: Metzler-Poeschel.

SRU (1998) *Gesamtgutachten "Umweltschutz: Erreichtes sichern - neue Wege gehen"*. Stuttgart: Metzler-Poeschel.

Stucke, A. (1993) *Institutionalisierung der Forschungspolitik*. Frankfurt: Campus.

Umweltbundesamt (1997) *Jahresbericht 1997*. Berlin.

Umweltbundesamt (ed.) (1998) *Innovationspotentiale von Umwelttechnologien*. Heidelberg: Physica.

VDI (ed.) (1991) *Integrierter Umweltschutz*. Düsseldorf: VDI-Verlag.

Victor, D.G. et al. (eds.) (1998) *The implementation and effectiveness of international environmental commitments: theory and practice*. Cambridge (Mass.): MIT Press.

Wallace, D. (1995) *Environmental policy and industrial innovation*. London: Earthscan.

Weidner, H. (1992) *Basiselemente einer erfolgreichen Umweltpolitik*. Berlin: edition sigma.

Wolf, R. (1988) "Herrschaft kraft Wissen" in der Risikogesellschaft. *Soziale Welt, 39*: 164-187.

Wolf, R. (1992) Sozialer Wandel und Umweltschutz. Ein Typologisierungsversuch. *Soziale Welt*, 43: 351-376.

LIST OF INSTITUTIONS AND THEIR ABBREVIATIONS

AfAS: institute for applied systems analysis (part of the research centre Karlsruhe) (research and policy consulting)

ATA: academy for technology assessment Baden-Württemberg (research and policy consulting)

BMBF: federal ministry for education, science, research and technology (since 1994)

BMBW: federal ministry for education and science (1969-94)

BMFT: federal ministry for research and technology (1972-94)

BMI: federal ministry of interior

BMU: federal ministry of the environment (since 1986)

DFG: German research society (organisation funding basic research)

DLR: German center for air and space research (research project management body for labour, environment and health research, among others environment-oriented technology and ecological systems research)

FhG: Fraunhofer Society (umbrella organisation of applied research institutes)

FZJ: research center Jülich

FZK: research center Karlsruhe

GSF: Society for radiology (research project management body for environmental and climate research)

ISI: institute for systems and innovation research (research and policy consulting)

MPG: Max-Planck-Society (umbrella organisation of basic research institutes)

Öko-Institut: research institute of environmental critics (research and policy consulting)

SFK: commission for technical incidents and accidents

SRU: expert council for environmental questions

TAA: council on safety of technical installations

TAB: office for technology assessment of the German parliament (research and policy consulting)

TUB: technical university Berlin

TÜV Rheinland: technical control association Rheinland (research project management body for earth-bound transportation technologies)

UBA: federal environmental agency (formally subordinated to the BMU; research project management body for waste management)

UGR: advisory board on total accounting in terms of environmental economics

UMK: conference of environmental ministers (of German states)

VDI: association of German engineers (involved in technical standard setting)

WBGU: scientific advisory board on global environmental change

WZB: science centre Berlin (research and policy consulting)

ZEW: centre for European economic research (research and policy consulting)

Chapter 5

Environmental Policy and Environment-oriented Technology Policy in the Netherlands

PETER S. HOFMAN AND GEERTEN J.I. SCHRAMA
Center for Clean Technology and Environmental Policy, University of Twente, the Netherlands

5.1   Introduction

This chapter presents an overview of the Dutch environmental policy (EP) and environment-oriented technology policy (ETP). The main aim is to give insights into some of the recent modifications and innovations in both EP and ETP systems, and to trace some of these changes back to the historical roots of EP and ETP in the Netherlands. The paper starts with an overview of environmental policy in the Netherlands, and a description of the standard EP system. This is followed by an introduction to recent developments in Dutch environmental policy. These involve the system of environmental planning, the focussing on target groups, the use of negotiated agreements, and stimulating the introduction of environmental management systems in companies. The next section focuses on environment-oriented technology policy. An overview of the development of ETP is provided by evaluating the main policy documents. Following this, the standard ETP system, the technology subsidy scheme, is reviewed. The main institutions and instruments are described in section 5.2 and the implementation of technology policy at the national and regional levels is explained. A number of new developments are then presented. The paper concludes with a review of the inter-policy co-ordination between EP and ETP in the Netherlands.

125

Geerten J.I. Schrama and Sabine Sedlacek (eds.) Environmental and Technology Policy in Europe.
Technological innovation and policy integration, 125-162. © 2003 Kluwer Academic Publishers. Printed
in the Netherlands.

5.2   Environmental Policy

5.2.1   The Standard EP system in the Netherlands[1]

5.2.1.1   History: direct regulation through licensing schemes

As early as 1810, when the Netherlands were under Napoleonic rule, a licensing system was introduced aimed at controlling the hazards, damage and nuisance caused by industry (De Koning, 1994: 13). The first environmental law was the 1875 Factory Act. Its successor, the 1896 Nuisance Act, which governed hazards and nuisance caused by installations at specific locations, remained the most important environmental law for a long period.

Environmental awareness grew in the late 1960s and early 1970s. Major landmarks were Rachel Carson's book 'Silent Spring' (1962), and 'Limits to Growth', the first report for the Club of Rome (Meadows, 1972). These made people aware of the deplorable state of the ecology and the finite availability of natural resources. In response to this, all over the world, environmental laws were put in place. In the Netherlands, as in many other countries, emphasis was put on prevention - to avert further decay. Activities and products harmful to the environment were either prohibited or made subject to licensing schemes. Reparation of existing damage was left to ad-hoc measures and the recuperative capacity of nature.

In 1962, 'environment' as a policy issue made its first appearance. In that year, the Minister of Social Affairs and Public Health set up a Public Health Inspectorate to be responsible for environmental protection. However, it was not until 1971 that environmental protection was established as a formal policy field when the Directorate General for Environmental Protection (DGEP) was created, as part of the newly established Ministry of Public Health and Environment. The DGEP was initially allocated only very limited resources. As the workload increased and environmental problems gained in urgency, the political will grew, notably in Parliament, to provide the DGEP with further resources and powers. The allocation of financial and human resources showed a marked increase in the period from 1972 to 1982. In 1982, the DGEP was transferred to the new Ministry for Housing, Spatial Planning, and the Environment (VROM).

Along with these institutional developments, a set of environmental laws was put in place. The dominant perspective was problem oriented. Individual environmental problems were consequently dealt with by integrating them into existing laws, often the Nuisance Act, or by enacting special laws, the so-called 'sector laws'. The latter include acts on: Surface Water Pollution (1969), Air Pollution (1970), Hazardous Waste (1976), Waste Materials

---

[1]   This section is mainly based on Bressers and Plettenburg (1997).

(1977), Noise Nuisance (1979), and Soil Clean Up (1982). In drafting the system of environmental laws little attention was paid to uniformity and internal coherence. This deficiency has led to sharp criticisms from various sections of society. Citizens have claimed that the public participation and appeal procedures were biased against them. Industry, for its part, claimed that the licensing procedures were far too time-consuming; companies needed a multitude of environmental permits, issued by various authorities who often imposed different and conflicting conditions (De Koning, 1994: 168-169).

To solve these problems, the 1979 Act on the General Provisions for Environmental Protection (in Dutch: Wabm) was introduced. This framework law subjected the various sector laws to uniform rules for the application and granting of licenses while also providing for uniform participation and appeal procedures. Nevertheless, it was still felt that environmental legislation lacked coherence, and pressure was exerted to further expand the scope of the new General Provisions Act at the expense of the sector laws. This culminated in 1993 with the incorporation of the General Provisions Act into a new Environmental Management Act (EMA, in Dutch: Wm). The name was deliberately changed to reflect that the new act was designed to create an all-embracing framework law. The EMA also opened up the possibility of granting integrated environment licences covering all environmental aspects. As a consequence, the five licensing systems stemming from the five environmental acts - the Nuisance Act, the Air Pollution Act, the Noise Nuisance Act, the Waste Materials Act, and the Hazardous Waste Act - were transferred to the Environmental Management Act as of March 1st 1993. This has made environmental legislation much more transparent and easier to enforce. A separate licence is however still required under the Surface Water Pollution Act, within the competence of the Ministry of Transport, Public Works and Water Management, since this ministry was unwilling to transfer this task.

5.2.1.2   Institutions and key actors

Policy institutions. To obtain a clear understanding of the policy institutions in the Netherlands, a subdivision has to be made using the vertical and horizontal distributions of power between the policy institutions. Horizontal distribution involves powers being divided between different authorities on the same level, such as Ministries. Vertical distribution relates to the division of powers between authorities at various government levels.

In the Netherlands, the administrative structure comprises three layers of government, namely: (1) municipalities and water boards, (2) provinces, and (3) the national government. These do not operate in isolation, but are complementary to each other. Initially, environmental regulations were exclusively made at the local level, and municipalities are still the most important government body for issuing and monitoring environmental licences. In the environmental laws of the 1970s, the provinces were given responsibility for their implementation. They also became the legal authority - i.e. for issuing licences, monitoring and enforcement of the law - for large, technically complicated, and highly polluting companies, a situation that has been confirmed by the Environmental Management Act. The national government concentrates primarily on national legislation and regulations as well as on the planning of national environmental policy, including the setting of targets and norms.

This, however, does not mean that the national government has sole responsibility for determining an environmental policy which municipalities and provinces are obliged to implement. The provinces, and most municipalities, conduct their own environmental policy planning. Moreover, the municipalities and provinces enjoy autonomous status when it comes to environmental policy. The law states that plans made at the various government levels are not governed by a hierarchical order. Vertical co-ordination should be achieved by means of mutual consultation, agreements, and the exchange of information. Where this fails, both the national government and the provinces have instruments at their disposal to enforce vertical fine-tuning.

The Dutch environmental policy planning system differs from systems in other countries, although some are similar to the Dutch example, by two characteristics. Firstly, it can be considered as an attempt to apply the ideas of the 'strategic choice approach' as it is known in the scientific planning literature. This means that the main goal is not to stipulate future actions, but to improve the coherence, the perspective on the future, the quality of motivation, and the openness for external participation in future decision-making (Coenen, 1992). The second characteristic, more or less in contradiction to this general orientation, is that the national, as well as many of the provincial and local, plans contain precise and quantitative targets, for instance on the decrease of emissions.

5.2.1.3   *Policy integration: inter-policy co-operation as an objective*

It has been argued in the Netherlands that the best environmental policy would be to put an end to all environment-unfriendly policies (Drees, 1992). This is a pointed reference to, among other things, the policy aimed at stimulating intensive agriculture and livestock farming, which involves the construction of large infrastructural works to increase the mobility of people and goods, land-use revision, and a lowering of the ground water level to facilitate agricultural mechanisation at the expense of nature.

A further factor is that many environment-related tasks have been entrusted to other Ministries. The Ministry of Economic Affairs, for instance, is responsible for energy-saving. The tasks of the Ministry of Agriculture include nature conservation, fisheries, and - together with the Ministry of Environment - manure disposal and pesticide use. The Ministry of Transport, Public Works, and Water Management is responsible for water quality management and also for curtailing the growth in car use. All the National Environmental Policy Plans were therefore drafted under the responsibility of all four Ministries involved. The coming about of the first National Environmental Policy Plan (NEPP) in 1989 can, incidentally, be seen as a successful attempt by the Ministry of Environment to take advantage of the wave of sympathy and attention for environmental issues in the late 1980s in order to secure the commitment of the other relevant Ministries to environmental objectives. This process was certainly not free of tensions. Shortly afterwards the government (Lubbers II, 1986-1989) resigned when the Ministers of Transport, Public Works and Water Management, and Environment, lost the support of the conservative Liberal Party after seeking to abolish tax relief for commuting traffic in an effort to slow down the growth in automobile use. In the framework of the NEPP, municipalities and provinces have been entrusted with all sorts of tasks designed to make the environment an integral concern in other policy fields. Their environmental policy plans must explain how this aim is given concrete form. This approach has also resulted in the creation of 'bridgeheads' at other Ministries, in the form of individual officials or complete departments which stand up for environmental issues within their own policy fields.

5.2.2   Transition to the new EP system

This 1980s are still seen as the major watershed in Dutch environmental policy. The transition to the present EP system is often attributed to Pieter Winsemius, Minister of Environment in the first Lubbers' administration (1982-1986). The shift can be characterised by three elements. Firstly, the new policy was based on the idea that environmental protection and the striving for sustainable development were not only the responsibility of government, but that each sector in society should take responsibility for solving the problems it creates. Secondly, as a consequence, the burden on the state, with its lack of problem-solving capacity, is reduced. Thirdly, it was expected that polluters who participated in solving their own problems would do so more diligently and also more efficiently than under a command-and-control approach. This approach fits quite closely with the broader strategies of deregulation and political modernisation pursued by the Lubbers' administration (Mol et al., 1998: 60).

The buzzword for the new EP system is 'internalisation', which means that everyone, every sector of society, assumes responsibility for the environment and gears its behaviour towards it; while the role of government is to set overall targets, to facilitate and to stimulate (see VROM, 1989a). It is a gradual process, characterised by subsequent stages; *"the wipe, the carrot, and the sermon"*, as Winsemius puts it (interview in Schrama and Van Lierop, 1999). Le Blansch (1996) stresses that internalisation, as intended in the Dutch environmental policy, is foremost a matter of standards and values, or civilisation in a sociological sense, and not the internalisation of external costs in an economic sense.

In the remainder of this section the most relevant aspects of the new EP system, from the perspective of industry, will be discussed: the changes in policy style, the national environmental policy plans, the target group approach and the ensuing negotiated agreements, the use of economic policy instruments, and the stimulation of corporate environmental management.

5.2.2.1   Policy style

Dutch environmental policy has gone through a radical change in policy style. Initially there was a wide discrepancy between the policy style of the policymakers and that of the policy implementers. The policymakers, civil servants and administrators at the Ministry of Environment, had adopted a distant, and often rather negative, attitude towards the regulated sectors such as industry. The addressees generally had little influence over the environmental policy, which tended to favour direct regulation. The enforcement of these regulations, however, was left to other agencies (Bressers, 1993; Bressers, Huitema and Kuks, 1994). The policy implementers, such as the licensing agencies, were understaffed. In addition, they often lacked motivation, and received too little support from the responsible administrators. As a result, they tended to respond to complaints rather than pursue active and systematic execution of the legal regulations. Systematic control and enforcement was virtually non-existent.[2]

The difficulties in providing concrete evidence of environmental infringements was a further reason for refraining from legal action and, instead, to opt for 'talk, talk, and more talk'. In other words, although the legislation was formally strict, in practice it was at best used as an informal bargaining counter in negotiations.

In the first half of the 1980s, Winsemius, the Minister of Environment, initiated vigorous efforts to persuade the environment policymakers and regulated sectors to abandon their entrenched positions (Winsemius, 1986: 61-67). He encouraged policymakers to see the policy addressees as 'target

---

[2]   Policy formulation based on consensus was already applied in the 1970s (e.g. Mol et al., 1998: 46).

groups' with which they could communicate. As this process gained momentum, special 'target-group managers' were appointed at the ministry. Their task was not only to act as ambassadors of the environmental policy vis-à-vis the target groups, but also to 'educate' their own organisation so as to have a better understanding of the target groups' positions. Contacts between environmental policymakers and the representatives of the target groups became much more regular. It was due to these initiatives that, at the end of the 1980s, the authoritarian style of Dutch environmental policymakers was supplemented with a new approach designed to encourage self-regulation. At the same time, the allocation of additional funds reinforced the policy implementation powers of environmental authorities at provincial and municipal levels, while growing public attention to environmental problems had also resulted in stronger administrative support for implementing the environmental laws. As a consequence, some friction has emerged between the implementers' regulation-oriented policy style, and the policymakers' emphasis on the companies' goodwill.

This problem has been recognised over the past few years, and scope has been created within the law to attach more weight to the companies' own environment plans in the licensing process. Further, frameworks have been created so that temporary 'toleration' of infringements is now subject to clearer rules. However, there is still a long way to go before the problem of opposing policy styles is resolved.

5.2.2.2   *National Environmental Policy Plans*

The headlines of the Dutch environmental policy are laid down in National Environmental Policy Plans (NEPPs). There is a certain pattern in these policy documents. The core of the present EP system was drafted in the first NEPP (VROM, 1989a) and elaborated in the Annex to NEPP 1 (VROM, 1990) and NEPP 2 (VROM, 1993). Subsequently, the emphasis changed, from policy formulation and acquiring support, to policy implementation and environmental management. This was expressed by Minister De Boer when she took office in 1994, and was reflected in NEPP 3 (VROM, 1998a) and NEPP 4 (VROM, 2001). Another development was environmental integration, the introduction of environmental concerns into other policy fields. Whereas NEPP 1 was drafted by the Ministry of Environment alone, NEPP 3 was published under the joint responsibility of the Ministries of Housing; Spatial Planning and Environment (VROM); Economic Affairs (EZ); Agriculture, Nature Management and Fisheries (LNV); Transport, Public Works and Water Management (V&W); Finance; and Foreign Affairs.

Briefly, the NEPPs are framed as follows (De Moel et al., 1999):

- explicit perception of the environmental problems (scales, material chains);
- policy approach (four-year planning cycle, two-track policy aimed at sources as well as effects, environmental themes instead of environmental sectors, target group approach, and an additional regional approach;
- distribution of responsibilities (central: policy targets and framework; decentralised: implementation).

Since the revision of Dutch environmental policy in the first NEPP of 1989, instruments have either appeared or gained dramatically in importance. These are aimed at stimulating sectors of society to accept responsibility for sustainability and environmental quality. This is done, for instance, through target group consultations and negotiated agreements, extended liability for environmental damage, tradable emission rights, research and information obligations, regulations requiring companies to employ staff with adequate expertise, and the creation of institutional facilities (environmental impact assessments, company environment departments, and internal company environmental management systems). Many of these instruments also operate indirectly through intermediary organisations, and sometimes even lead to the creation of such intermediary organisations. This approach has become much more widespread, not only in the Netherlands but - since the fifth Environment Action Programme of the European Union - also at the European level (Bressers and Plettenburg, 1997).

5.2.2.3   Target groups[3]
In the 1989 NEPP 1, the so-called 'target group approach' was introduced. Eight social sectors, responsible for considerable environmental impacts, were labelled as target groups. The government started talks with their representatives in order to establish joint policies for the control and reduction of their impacts. The selected target groups are:
- agriculture,
- transport and mobility,
- industry,
- energy production,
- oil refinery,
- construction,
- waste processing,
- consumers and retailers,
- drinking and wastewater (added in NEPP 2).

---

[3]   This section is mainly based on Bressers and Plettenburg (1997).

In an attempt to encourage the internalisation of environmental responsibility by individual actors a consultative structure around government and industrial organisations, representing the target groups, has been created. One of the main aims of the consultation process is to define the tasks for a specific sector within the framework of the overall national environmental objectives. Experience from the first years has shown that target groups consisting of large numbers of small units (such as households, farmers, car drivers, retailers, and other small firms) are the most difficult to reach by means of a target-group policy (NEPP 2).

5.2.2.4   Negotiated agreements

Negotiated agreements - also called voluntary agreements or covenants - are the most common kind of policy instrument in the Dutch policy approach that draws on consensus and self-regulation. Negotiated agreements are formal agreements between government and representatives of societal sectors - usually branches of industry - in which the latter make certain commitments, often as an alternative to direct regulation. Most negotiated agreements concern issues that are assumed to be the main environmental problems related to the sector concerned. Several classifications can be made, such as waste reduction (collection of used batteries, reduction of packaging materials), reduction of emissions of specific substances or classes of substances (VOCs, CFCs), and integrated negotiated agreements concerning all the environmental effects of particular sectors or branches. In addition to the target-group covenants, there are also energy conservation agreements between the Ministry of Economic Affairs and the industrial sectors.

The application of negotiated agreements is mainly based on trust between the parties. They have no foundation in public law. It is generally assumed that they constitute contracts under private law, but this may not always be the case (Algemene Rekenkamer, 1995), and it is not certain whether compliance can be enforced through the courts (Mol, 1998: 73). We are not aware of any relevant jurisdiction.

The negotiated agreements, as part of the target-group approach for industry, are framed in a six-step scheme, with a four-year cycle:
1. Firstly, emission reduction targets were formulated for industry as a whole.
2. Next, fifteen priority branches of industry, involving 12,000 companies responsible for over 90% of industry-based environmental pollution, have been selected by the Ministry of Environment.
3. Negotiations were then started with each of the fifteen branches. The Ministry of Environment and the trade associations established the so-

called 'Integral Environmental Targets', mainly in terms of emission reductions, at the branch level. Targets were set for 1995, 2000, and 2010 relative to a base year that varied depending on available emission data. The outcomes were recorded in negotiated agreements that were officially signed by all parties. Large companies and representatives of the regulators (provinces, municipalities, and waterboards) were also involved. Part of the deal was that the branch would not be affected by new legislative measures, but that the environmental permits of participating companies would be adjusted to reflect the content of the agreement.

4. The Government facilitated extensive information campaigns aimed at individual firms, which were carried out by an independent agency (FO Industrie), with a considerable stake for the trade associations.

5. For the implementation of the negotiated agreements, a distinction has been made between homogenous and heterogeneous branches (according to firm size and production processes). For the former cases, an 'Environmental Handbook' or 'Manual' was developed, which served as a uniform directive for each participating company. In the latter case, each participating company was expected to develop an individual 'Corporate Environmental Plan' which would be submitted for approval to the main regulator (either the municipality or the province). Corporate environmental plans would be renewed every four years, and based on assessments of the state-of-the-art of the technologies in the respective branches. New technologies which are accepted as applicable to a certain sector are labelled as 'specific measures' and should be incorporated when corporate environmental plans are renewed. In the case of the metalworking and electrical engineering industries, however, companies develop corporate environmental plans based on general directives (technological state-of-the-art for different processes) as laid down in a guidebook.

6. The final step involves implementation within the practice of corporate environmental management, and the adjustment of the firm's environmental licences. With respect to monitoring and control, participating companies have to submit annual progress reports, according to a framework that is laid down in the negotiated agreements. The issue of reporting is backed up by a law on environmental reports which came into force in 1999 and affects the 300 largest and most polluting companies.

The process was subject to considerable delays. Only three negotiated agreements were in place on schedule (i.e. by the end of 1993). These concerned the chemical industry, the basic metals industry, and the printing industry. However, these were also the most important ones since these industries were responsible for 60% of the total industrial environmental deterioration. In two instances, the negotiations were still ongoing (as of August 1996), and in six cases the parties involved agreed not to have a covenant. Table 5.1 gives an overview of the branch agreements currently in operation.

Table 5.1: Overview of Target Group negotiated agreements for industry

Industry	Companies[*]	Type	First agreement
Primary metals industry	39 (38)	Heterogeneous	1993
Chemical industry	148	Heterogeneous	1993
Printing industry	3400	Homogenous	1993
Dairy industry	34	Heterogeneous	1994
Electroplating industry	±18.000	Hybrid	1995
Textiles processing, carpet and floor coverings industry	80 (49)	Heterogeneous	1995
Paper (products) industry	28	Heterogeneous	1996
Concrete and cement industry	440	Homogenous	1998
Rubber and plastics processing industry	>1.200	Homogenous	2000
Meat industry	332 (185)	Homogenous	2000

* Number of companies in the branch; between brackets number of companies as signatories to the negotiated agreement. Source: FO Industrie (www.fo-industrie.nl), last update October 2001.

All the requirements for individual firms are extensively described, as well as the sanctions to be applied to negligent firms, in the event of total refusal or sluggish behaviour. Although certain firms have refused to endorse the covenants, there is no evidence that they perform significantly less that other firms. In general, sluggish firms are treated very considerately with gentle reminders and neatly-formulated requests.

In general, all parties are very satisfied with this approach. There is, however, a tendency to downplay certain problems:

- the Corporate Environmental Plans are rather superficial and show no profound vision on corporate environmental management;
- the guiding principle of 'best available technology' is very difficult to apply in specific cases;
- the target group approach by individual branches of industry disregards product chain interdependencies;
- monitoring and control by environmental regulators is very demanding, the agencies concerned often lack adequate expertise and manpower;
- monitoring by environmental regulators is usually not very precise, since they are very careful not to disrupt the good atmosphere;
- many companies have difficulties in complying with the requirement for annual progress reports, as this has turned out to be a demanding task;
- the consultations are confined to two kinds of actors, government and industry; while third parties, such as environmental groups, are not

involved (environmental permits are based on negotiated agreements and corporate environmental plans, and third parties have the right to challenge them in court, but in practice the chances of success for this kind of procedure are very low and this option is rarely pursued - Biekart, 1995).

Government and industry are clearly very pleased with the results thus far. There is no doubt that things have changed and the relations between government and industry have improved. There is no doubt either that industry is active with the implementation of environmental management and the public advertising of their achievements in this field. The environmental movement, however, has strong doubts as to whether companies, in particular large ones, are doing any more than what they would have done under the old regulatory regime.

5.2.2.5   *Economic instruments*
Economic instruments are not widely applied in the Dutch environmental policy. The discussion on this topic started in the first NEPP of 1989. There have been many debates, with very few outcomes. In 1995, the OECD reviewed the Dutch environmental policy very positively, but it judged that emphasis should be put less on consensus building and covenants, and more on economic instruments (Opschoor, 1995). When the Kok administration took office in 1994, it presented the 'greening of the tax system' as one of its priorities. A commission headed by Van der Vaart was charged with the task of investigating possibilities and barriers. At the start of its second term, in 1998, the Kok administration announced additional measures aimed at a gradual shift from taxing of labour to taxing of environmental deterioration. An important role was here assigned to the new system for income tax, taking force in 2001. This section will now dwell somewhat more on the topics of green taxes and levies, and stimulating fiscal measures. Other economic instruments – not discussed in further detail here - are the tradable quota in the field of agriculture (for manure and milk production) and fisheries. Although these measures are, at least to an extent, implementations of EU policies, they do have considerable effects on the behaviour of the target groups.

Green taxes. The difference between taxes and levies is that the revenues from the former are part of the general means, while the revenues of the latter are earmarked for purposes related to the activities on which they are imposed. In practice, however, the distinction is somewhat blurred. In official government documents (e.g. VROM, 1999) the following 'green taxes' are distinguished:
– excise duty on mineral oils (the regular tax on lubricants);

- excise duty on the sale of motor vehicles and the annual duty on the use of motor vehicles;
- environmental taxes.

Total expected revenues for 1999 were NLG 26 billion (€11.8 billion), about 14% of the total tax income. The definition of 'green tax' is admittedly very broad and includes regular taxes and duties related to automobile usage. There are five 'real' environmental taxes with revenues of about NLG 5 billion (€2.3 billion) in 1999. These are[4]:

- fuel: an environmental tax on fossil fuels in addition to the regular excise duty; the tax is levied primarily on the manufacturers and importers of fuel;
- groundwater: tax on the extraction of fresh water, introduced in 1995, with rates at about €0.18/m³ when used for drinking water, and €0.09/m³ when used for agricultural and industrial purposes;
- waste disposal: general levy on the delivery of waste on top of the price of waste processing; specific types of waste disposal are free of charge, such as organic household waste, and asbestos;
- regulatory energy tax: levied on the use of natural gas, electricity, and mineral oils used for heating by households and small companies (see below for more detail);
- uranium: tax levied on energy companies to counterbalance the taxes on fossil fuels.

Energy taxes. Attempts to introduce an energy or carbon tax have failed several times, mostly due to strong resistance by industry, certain parties in parliament, and the Ministry of Economic Affairs. When, in 1994, the Kok administration took over, a new opening was created and consensus was reached on a restricted and small energy tax for households and small companies, which was finally introduced in 1996. In order to prove that the measure is aimed at the reduction of energy consumption and not at increasing taxes, the revenues are returned by way of reductions in the rates of other taxes. Extension of this measure to larger companies is dependent on developments within the European Union, in order to avoid competitive disadvantages for Dutch industry.

Environmental levies. There are several levies charged for environmentally burdensome activities. The most important ones are: the sewage rights, waste levies, and the wastewater levy for discharges into surface waters. Total revenues for 1999 were expected to be about NLG 6.8 billion (€3.1 billion). The levy on wastewater is discussed in more detail below.

---

[4]   Source: website Ministry of Finance (www.minfin.nl).

Wastewater levy. Notwithstanding the limited application of economic instruments in environmental policy, the Netherlands are well known for the early implementation of a wastewater levy. This levy, introduced in 1970, has proved to be successful both in generating funds for the setting up and maintenance of wastewater treatment plants, and in curbing wastewater production by industry. Whereas industrial production increased steadily from 1970 to 1980, organic wastewater pollution by industry decreased dramatically in the same period (Bressers 1983; 1988). The introduction and development of pollution charges was found to correlate strongly with the reduction in wastewater pollution[5] (Bressers, 1988: 507). The high level of charges, providing a significant incentive for companies, has explained the relative success of the wastewater levy in the Netherlands compared to other countries. As a comparison, annual per capita revenues from water effluent charges in the mid-1980s were about €37 in the Netherlands as against €4.3 in France and €2.2 in Germany (Opschoor and Vos, 1989).

Fiscal measures. The Ministry of Finance has a pretty long list of fiscal measures, but most of them have had limited impacts. The most important ones are: a tax deduction for investments in energy efficiency (EIA), a tax deduction for environmental investments (MIA), a tax depreciation measure for investments in environmental technology (VAMIL), and a measure on 'green investments'. The latter two green tax measures are well known success stories of the last decade. Both are linked to deputy-minister of Finance, Willem Vermeend, and will be discussed briefly below.

VAMIL measure. As a Member of Parliament, Willem Vermeend initiated a tax depreciation measure for investments in environmental technology, the so-called 'VAMIL measure', which was introduced in 1993. The basic idea is that companies have the maximum leeway in the depreciation of specific equipment (the present list includes about 480 items), and this can yield considerable tax advantages. There are also less generous tax measures for regular energy and environmental investments.

Fiscal measure on 'green investments'. In 1995 Vermeend introduced another of his ideas, a fiscal measure on 'green investments'. The measure involves an exemption from an individual's income tax of interest and dividends earned from approved green investment funds. The measure is aimed at generating money at favourable interest rates for sustainable projects in the fields of energy, nature, and agriculture. The measure is considered to be a great success. Within a few years all major banks had introduced their own 'green investment fund'. On its introduction, there was

---

[5] The extent and change in pollution charges varies from region to region in the Netherlands, Bressers found a significant relationship between decrease in wastewater pollution and increase in pollution charge by differentiating between industries and regions.

a run on these funds and banks were faced with a temporary shortage of eligible green projects. In 1996 and 1997, the Government expanded the criteria for eligible projects. The number of issued certificates for green projects has risen from 160 in the first year to 539 in 2001. The total value of these projects was €336 million in the first year and €914 million in 2001 (Novem, 2002).

5.2.2.6   Self-regulation through stimulation of environmental management in companies

As explained earlier, the first NEPP introduced the concept of self-regulation as a cornerstone of environmental policy. It was concluded that directives from government actors would not be sufficient to realise the changes necessary for adopting a path towards sustainable development. Furthermore, the environmental problems of industries were acknowledged to be too complex and too dynamic to be solved by static regulations such as environmental standards. Therefore, a new policy approach was developed with a basic premise that companies themselves are able to clean up the environmental problems they create, and secondly that companies can be made aware of the environmental problems which they cause. In a policy document (VROM, 1989b), therefore, a strategy was set out in which industry would be induced to set up environmental management systems (EMS) within companies. These systems would provide the tools for companies to gain insights into their environmental impacts, and provide tools to control and decrease that environmental impact. The government expected, according to the policy document, that companies with large environmental impacts (estimated number about 10,000) would have a functioning environmental management system in place by 1995. Smaller companies were expected to install relevant parts of an environmental management system by the same time. Related to this, it was also expected that government organisations, such as municipalities and provinces, would implement environment management systems. Given the ambitiousness of these goals, a specific stimulation policy was implemented. The main strategy was to have intermediaries that could convince companies of the importance of environmental management (De Bruijn and Lulofs, 1995). It should be noted, however, that, independent of these government initiatives, employers' organisations and branch associations were already active in developing tools for environmental management for their members (De Bruijn and Lulofs, 1996). The government strategy consisted of a programme of activities with four clusters, and a total budget of around €23 million (De Bruijn and Lulofs, 1996: 48-49).

Research has shown that branch associations have played an important role in informing and supporting companies concerning the importance of environmental management systems (De Bruijn and Lulofs, 1995). For most branches, handbooks on the implementation of EMS in companies have been developed. In 1996, more than one-third of all Dutch companies had a functioning EMS, while another 37% had implemented several elements of an EMS as defined by the government policy document. The percentage of larger companies with implemented EMSs is considerably higher than that for SMEs.

With the introduction of certification schemes such as BS 7750, and later ISO 14001, several Dutch companies applied for certification for their EMS. In 1998, more than 200 companies had an EMS certified according to ISO 14001, and this figure is expected to double annually. EMAS, however, is much less popular with Dutch industry, with only around ten companies certified in 1998. Some of the reasons given are that EMAS is not a worldwide certification scheme that the required certified annual environmental report does not have any added value, and moreover that companies would be reluctant to publish it for competitive reasons (Van Oorschot, 1998).

5.2.3   Evaluation of Dutch environmental policy

A positive feature of Dutch environmental policy is the general framework laid down in the National Environmental Policy Plans. The general picture of a worrisome environmental state has generated a basis for environmental policies in various government departments, in business, and in the public at large. Moreover, the goals in the policy plans have made clear that there is more to do than 'business as usual'. Direct regulation, although effective in curbing obvious pollution, has not been effective in realising more prevention-oriented behaviour in business and society. The wastewater levy has been one of the most successful examples of Dutch environmental policy, but has not led to an extensive use of economic instruments in, for example, energy policy, mainly due to their alleged negative impact on competitiveness. Environmental management systems have now become widespread in Dutch companies, but this does not necessarily imply a significant reduction in environmental impacts. Negotiated agreements have become increasingly popular in Dutch environmental policy making, but the results of these policies are difficult to assess. While being positive about the co-operative mode of governance, in which various actors take responsibility for ensuring the environmental changes deemed necessary take place, it is unclear, and somewhat unlikely, that they result in the kind of change necessary for sustainable development. Table 5.2 summarises some of the results of negotiated agreements. Furthermore, we give an assessment of the potential of the new approach for delivering the kind of changes that are needed to achieve sustainable development. The success of a negotiated agreement correlates positively with the degree of organisation of the target group. This indicates that in the case of heterogeneous and unorganised target groups it is much more difficult to make progress through the negotiations. The negotiated agreements allow target groups to time the development and implementation of measures, but it is not clear whether this will result in changes that are more radical in the longer term.

Table 5.2: Results of negotiated agreements and their possible contribution to sustainable development (source: Hofman and Schrama, 1999)

Dimension	Results	Possible contribution to sustainable development
Degree of freedom of choice	Target groups can time their own implementation of measures to some extent, however 'alara' ('as low as reasonably achievable') is the bottom line for individual companies.	It is not clear who will take the initiative to develop new technologies, the anticipated market is still the main factor for inducing innovation. Some of the choices for new technologies and especially products do not bring sustainable development closer.
Co-operative nature	Intensive negotiations between target groups and government: regular meetings improve acceptance and the basis for environmental policymaking.	Consultation takes place between existing companies with significant interests in the current modes of production and this may impede more radical innovations.
Level of ambition	Long-term targets (2010) are ambitious; short-term targets (1995) reflected state-of-the-art technology and tended towards 'business as usual'.	Depends on the realisation of necessary innovations. Current targets are mainly emission ones, and less focussed on input/resource use, whereas the latter is a condition for progress towards sustainable development.
Time horizon	Extension of the time horizon, especially for the mid (2000) and long-term goals (2010) provides direction and some certainty for industry.	Depends on the effectiveness of the links with technology policy; a lengthy period is needed to develop new technologies and product-consumption linkages.
Instrumentation	The mix of instruments works to some extent: most target groups are on schedule to meet most targets. The difficulty is how to switch to strict enforcement when targets are not within reach.	The mix of instruments needs to exert enough pressure and scope to drive and facilitate industries along a path towards sustainable development. Co-ordination between the various policies in different policy areas is needed.
Addressees of policy	Associations and active companies are being reached, but it is harder to reach sluggish/defensive companies. The success of the network approach depends on the degree of organisation of the target group.	Many of the goals of sustainable development require collaboration between various actors. Relations with consumptive and other industrial sectors is limited, while much progress might come from changes in these linkages or technological developments outside the branch.

In addition, the consensual nature of negotiated agreements, and the network of actors currently involved in the agreements, promotes innovations of an incremental nature rather than radical innovations, because newcomers, or new technologies which might be developed outside the branch, are not part of the agreement. More research is needed on whether the consensual and target group approach is conducive to innovation of a more radical kind. The diffusion of state-of-the-art technology, as propagated in negotiated agreements, will only be successful if it is complemented by direct regulation, which implies that when direct regulation fails, negotiated agreements are also likely to fail. The successful development of 'new' technologies will depend on whether negotiated agreements are complemented by programmes which specifically inform the target group about, and focus on, the technological paths and technologies which need to be developed in order to solve technological bottlenecks.

For the sake of sustainable development the linkages between production and consumption, and within product chains, are crucial and need to be reconsidered. However this is well beyond the scope of the current negotiated agreements and, in this sense, a new generation of negotiated agreements which capture these dimensions would perhaps have greater potential to facilitate progress by industry towards sustainable development.

As mentioned, the target group policy for industry had reached the implementation phase under NEPP 3 (1998). The four-year cycle has been institutionalised and was no longer subject of political debate. In general, the sense of urgency had subsided and the impression had grown that the big environmental problems had been solved. The 2001 NEPP 4 (VROM, 2001) was aimed at reformulating Dutch environment policy. It indicated seven most urgent environmental problems and indicated the paths to solve them in the period till 2030. The seven problems were:
– loss of biodiversity,
– climate change,
– over-exploitation of natural resources,
– threats to health,
– threats to external safety,
– damage to the quality of the living environment,
– possible unmanageable risks.

Since NEPP 1 (1989) the role of industry as the main target group of environment policy has gradually diminished. NEPP 4 distinguishes between problems related to the economic system and more complex ones. For the former, short-term thinking, fragmentation and institutional shortcomings - including government - and the fact that environmental costs are hardly included in the prices of products and services, are major obstacles for reaching sustainability. The more complex environmental problems require technological, economical, social-cultural, and institutional transformations that can only be achieved in the long run. These are called 'transitions', and NEPP 4 indicates the following transitions as the major challenges of the Dutch environment policy in the coming decades:
– transition to a sustainable energy system (emission, energy, and mobility),

- transition to sustainable use of bio-diversity and natural resources,
- transition to sustainable agriculture (environment, nature, and agriculture).

Although the role of industry has changed, it still will have an important stake in the solutions for these environmental problems.

5.3   Environment-oriented technology policy in the Netherlands

Based on the lessons learnt from the NEPP 1 and NEPP 2 policies, the Dutch government concluded that environmental policy should enter a new phase, that of 'environmental management'. In the previous periods, the main focus was on cleaning up pollution and a reactive attitude towards environmental problems was adopted. Nowadays, the main objectives are to ensure an absolute decoupling of economic growth from environmental pressure, and the sustainable use of resources (VROM, 1998a: 35). Technology is expected to play an important role in reaching this objective. According to the NEPP 3, *"science and technology will play a vital role in bringing sustainable development closer and in providing insight into the technological and social breakthroughs sought"* (VROM, 1998a: 41). As a result, the NEPP not only provides the agenda for environmental policy, it also sets the stage for environment-oriented technology policy in the Netherlands.

Another pillar of environment-oriented technology policy stems from the Dutch technology policy within the responsibility of the Ministry of Economic Affairs. In line with the NEPP perspective, which is also the responsibility of this ministry, it is argued that the challenge for economic policy is to ensure that the Netherlands can follow a sustainable path while achieving high economic growth. The stimulation of technology development and the provision of information and advice are important pillars of this technology policy. Furthermore, from 1994 on when the first Kok administration took office, the importance of innovation-oriented clusters has been emphasised and specific initiatives taken to stimulate the formation of clusters. Some of these initiatives will be explored more fully in the next section in describing the roots of Dutch environment-oriented technology policy and explaining its current focus.

5.3.1   A short history of ETP in the Netherlands

Technology policy became a separate policy field in the 1970s, when a comprehensive 'innovation policy document' was published (Tweede Kamer, 1979). One of the drivers of technology policy is the importance of competitiveness to the Dutch economy. Technology policy can play a role in improving the international competitiveness of Dutch industry, and government support for industry is partly legitimised using the 'level playing field' argument, referring to state support given to foreign competitors (Van Dijk and Van Hulst, 1988). Until the 1980s, technology policy was developed separately from environmental policy, without any focus on environment-oriented technology. At that time, the Ministry of Environment started a programme to stimulate clean technology in order to stop the spread of pollutants, but the focus of the actual projects was mainly on end-of-pipe technologies. This was also reflected in the environmental law and permits, which set short-term standards to reduce emissions. As is shown in Table 5.4, environmental problems were mostly dealt with by companies installing standardised end-of-pipe technologies. However, the increasing public attention to environmental problems, the worrisome conclusion of 'Concern for tomorrow', a report on the state of the environmental in the Netherlands by the national institute for public health and the environment (RIVM, 1988), and other factors, led to changes in both environmental and technology policies.

Table 5.3: Development of environment-oriented technology policy in the Netherlands from the 1970s onwards.

Period/characteristic	1970s to mid-1980s	Mid-1980s to mid-1990s	1990s
Focus of environment-oriented technology policy	Emission-oriented	Process-optimising	Prevention-oriented
Orientation of companies	Treatment of emissions	Reduction of emissions	Reduction of raw material usage
Nature of technology	End-of-pipe technology	Process-integrated measures	New processes and product chains
Focus of environment-oriented projects	Standardised for individual companies	Customised for individual companies	Co-operation between companies on industrial estates and in production chains

According to the 'Concern for tomorrow' report, emissions of several polluting substances needed to be reduced by between 70% and 90% in order not to overstress the capacity of the environment. The practice of adopting end-of-pipe measures were costly and did not lead to improved environmental performance. According to RIVM, a sustainable alternative needed to include: ongoing prevention, re-use, energy saving, and the development of the technology necessary to achieve this (RIVM, 1988; see also Cramer, 1996: 132). The first National Environmental Policy Plan reflected this shift in environmental policy and adopted the recommendations of the RIVM report. The main lines of the Dutch environmental technology policy were developed in a policy document on 'Technology and Environment', jointly drafted by the Ministries of Economic Affairs and Environment (EZ and VROM, 1991). This document reflects a change in policy that had previously been mainly concerned with remedial measures that tended to induce the use of end-of-pipe technology. The document further developed a policy path in which the dominant focus would be on preventing pollution at the source by implementing process-integrated and clean technology. This also implies a shift from a one-sided focus on the supply side of technology towards an orientation on the mutual adjustment of demand and supply of environment-oriented technology. Several programmes were developed, and others were adapted, in order to ensure the effective stimulation of technologies to solve long-term environmental problems, and to increase the users' appreciation of both the environmental and economic potential of new environment-oriented technologies.

A further shift towards emphasising the importance and scope of environment-oriented technology took place when the role of technology as a link between ecology and economy was advanced. The strategy document 'Knowledge in Action', a joint production by the Ministries of Economic Affairs, Education and Science, and Agriculture (EZ, OCW, and LNV, 1995), implemented the Dutch government's new policy initiatives to increase the knowledge-intensity in the Dutch economy, in order to enable it to move onto a higher and more-sustained growth path. By doing this, it was argued, the link between economy and ecology could also be strengthened, not just by tackling environmental problems, but also by making greater use of technological opportunities to create a sustainable economy. The document stresses the importance of environment-oriented technology as one of the spearheads of technology policy. Some specific actions were set out to advance these ambitious goals. One was the setting up of a new programme, 'Economy, Ecology, and Technology' (EET), with a focus on the development of technological breakthroughs that would provide a leap in efficiency and an improvement in competitiveness in the longer term.

In the 'Environment and Economy' policy document, which gave a perspective on sustainable economic development, some more-specific options and action paths were described (VROM, 1997). Various activities were explored which could result in so-called 'win-win situations', where both the environment and the economy would benefit. In current governmental policies, therefore, the main philosophy is that the strengthening and renewal of the economic structure can coincide with improved environmental and energy efficiency. The document reflects not only co-ordination of EP and ETP, but also steps towards their integration. In several actions, it is stressed that aspects of both regulation and technology have to be dealt with. An example is that of product improvement involving policy measures on ecolabels, product standards, and the stimulation of product-oriented environmental management; combined with technology programmes focussed on environment-oriented product design and chain management. A further example is the development of sustainable industrial estates where both regulatory actions and technological initiatives are required. To sum up, several initiatives have been set in motion where the Ministries of Environment and Economic Affairs jointly develop action paths towards a more sustainable economy.

5.3.2   Institutions and instruments for ETP

5.3.2.1   *Generic and specific subsidies for environment-oriented technology*

The Ministries of Environment and Economic Affairs are the most important actors with regard to environment-oriented technology policy. At the national level, the Ministry of Economic Affairs is primarily responsible for technology policy. Part of this policy is the establishment of a number of generic programmes under which subsidies and credits are applied for the development and application of new technologies. Several specific programmes are focused on the promotion of the development and application of energy-relevant technology. In addition, the Ministry of Environment has put in place several programmes on environment-oriented technology policy, as part of its environmental policy. Various other departments are involved in stimulating environment-oriented technology policy, in most cases in co-operation with either the Ministry of Economic Affairs or the Ministry of Environment. Although environment-oriented technology policy contains a mix of instruments, such as various financial incentives, information, and advice, applied by various agencies the national and regional governments, the dominant instrument is the subsidy scheme. The number of stimulating programmes focussing on environment and energy is very high; in 1998 about 150 were in operation (VROM, 1998b). Of these, about one-third was focused on energy savings, efficiency and renewable energy. The other two-thirds were focused on environmental improvements to products, processes and product chains. Approximately two-thirds of the programmes are implemented at the national level. Projects at the regional level mostly focus on providing information and advice to relevant organisations. Table 5.4 presents a selection of the main financial schemes employed to stimulate environmental innovations in Dutch companies.

STD programme. The largest technology programme to date with a focus on the environment was the 'Sustainable Technological Development programme' (STD, Dutch acronym: DTO), performed in the period 1993 to 1997, with an annual budget of around €45 million funded by five Ministries (Economic Affairs; Agriculture; Education and Science; Traffic and Water Management; and Environment). A large number of companies, research institutes and NGO's have contributed to the programme and the performance of the projects. The programme was based on a specific and extensive philosophy and method aimed at achieving long-term sustainable technological development. Some characteristic elements from the philosophy and method are (Weaver et al., 2000):

- Government should influence innovation processes indirectly by encouraging and facilitating the context wherein the processes take place in stead of 'picking and backing prospective technological winners'.
- Integration of environmental policy and environment-oriented technology policy.
- Innovators should attend and integrate technological, cultural and structural aspects.
- Social networks are key elements both in the stabilisation of present technologies and in the creation of new breakthrough technologies, so 'engage the stakeholders'.
- Long-term solutions determine the short-term steps to be taken. Therefore a time horizon of fifty years was taken. Technological improvements to be achieved, in terms of eco-efficiency, are estimated at a factor twenty. This factor is computed as the product of the present environmental pressure which is twice as high as it should be, the expected doubling of the population, and the envisaged increase of individual prosperity by a factor of five. The required short-term steps are establish by way of 'backcasting', that is taking the desired situation fifty years from now as point of departure.

To the STD programme five sub-programmes were defined, according to the 'key need areas' (Weaver et al., 2000: 78): water, building and spatial planning, transport and mobility, materials and chemistry, and nutrition. The sub-programmes were distinct entities within the programme with an assigned staff and a steering group. The task was to assess the needs for improvement of eco-efficiency on the long term for the specific field. In addition a number of projects on selected issues should be initiated with the aim of setting innovation processes in motion and engaging the relevant social parties.

The STD programme has been terminated ultimo 1997. Follow-up activities have been unfolded in a large variety. At the overall programme level, in September 1998 a new programme, 'DTO-KOV' has been established at a smaller scale, aimed at knowledge transfer and education.

Legal aspects. In terms of subsidies, the 1997 General Act on administrative law applies (AWB). This act provides a framework for the procedures of granting subsidies and carrying out the concerned regulations. Specific articles provide rules for the legal basis of subsidies. Legal regulations specifically prescribe the activities that are eligible for a subsidy, and procedures in terms of the obligations of the granting organisation and of the recipient. For specific programmes, rules and directives are laid out in specific documents which are published in the government's official journals (Staatscourant and Staatsblad). Two dedicated organisations, Novem (Netherlands Agency for Energy and the Environment), and Senter (Agency of the Ministry of Economic Affairs), are charged with the administrative and procedural tasks for these programmes.

5.3.2.2 Provision of advice and information through support for intermediaries

Apart from the programmes primarily focussed on stimulating the development of environment-oriented technology, the government also induces diffusion of innovations by establishing and funding intermediary organisations that inform companies about innovative options and companies involved in innovation networks. Figure 5.1 provides an overview of organisations in the Netherlands that are involved in the stimulation of development and diffusion of environment-oriented technology.

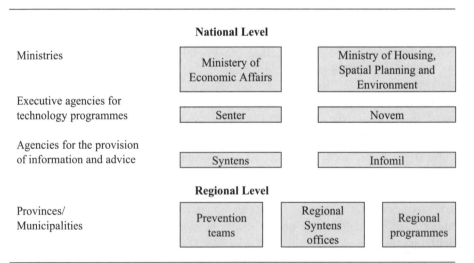

Figure 5.1: Organisations involved in the stimulation of development and diffusion of environment-oriented technology in the Netherlands

Table 5.4: Overview of financial schemes to stimulate environmental innovations in companies

Financial scheme Sponsor Budget 2002	Objectives	Specific characteristics
Economy, Ecology, and Technology (EET) EZ & OCW €22.7 million	Stimulating research into new technologies that may bridge the gap between ecology and economy.	Specific focus on technological breakthroughs and co-operation between users and developers of technologies. Significant results from the project are expected over 5 to 20 years. Subsidies up to €4.5 million per project are possible (25% - 60% of total project costs).
Subsidies for energy programmes (BSE) EZ €113 million	Support for projects aimed at: energy efficiency, use of renewable energy, and application of energy technologies with reduced environmental impact.	Framework measure consisting of several programmes focused on specific industries, or on the promotion and development of energy technologies based on renewable sources.
Fiscal depreciation of environmental investments (VAMIL) Finance & VROM €63,5 million	Stimulating investments in new machinery or equipment with less environmental impact than the existing ones.	Tax measure enabling companies to choose the moment of depreciation of investments in environment-beneign equipment (contained in a list of approved items).
Technology co-operation (BTS) EZ € 23.6 million	Stimulating co-operation in R&D projects by SMEs (up to 25.000 employees).	Criteria are a high level of innovativeness and good economic perspectives. Project needs to involve at least two partners.
Programme Environment and Technology VROM a.o. €2.6 million	Solving technological bottlenecks in industrial branches in order to realise the reduction targets in negotiated agreements.	Specific environmental problems are formulated within the programme, relatively high subsidy percentages, project information is public.
Innovation-oriented research programmes (IOP) EZ €18 million	Stimulation of co-operation between research and industry aimed at bridging the gap between fundamental research and need for applicable new technology. Environment-oriented technology is one of the foci.	Framework measure consisting of several subsidy programmes for universities and non-profit research institutes.
Technical development projects (TOP) EZ €25 million	Promotion of the development of products with reduced environmental impact. (successor to MPO)	Only applicable to SMEs: a project should result in a significant reduction in environmental impact relative to existing products. Credit is granted up to 40% of project costs and if the product fails to realise turnover within five years of its introduction the credit is written-off.

Sources: Arentsen & Hofman, 1996; Hofman, 1997; VROM, 1998b; Schrama, 2002; www.senter.nl.
EZ = Ministry of Economic Affairs; VROM = Ministry of Environment; OCW = Ministry of Education and Science; LNV = Ministry of Agriculture.

Further information is provided in the following section. Several organisations are involved in providing information and advice at the national level. An information centre, Infomil (the Dutch abbreviation for information point on environmental licensing), has been set up to inform government agencies, such as local environment agencies, companies, and citizens about current environmental standards, anticipated standards, and the current state of the art with regard to technology. Furthermore, the Ministry of Economic Affairs has set up a network of innovation centres, Syntens, regional offices in all provinces. Syntens mediates between companies and sources of knowledge, such as research institutes and universities. Syntens is regionally organised through fifteen establishments in the various provinces in the Netherlands. Syntens' activities are funded by the Ministry of Economic Affairs and other government agencies. Its major objective is to stimulate innovation of products, production processes, and management in Dutch companies through performing the following tasks:

–   provide advice to companies;
–   link company demand to the supply of knowledge;
–   guide innovation processes that companies carry out in co-operation with other organisations;
–   provide information to entrepreneurs and policymakers;
–   stimulate specific regional activities focussed on innovation in companies;
–   develop methods and tools in order to advise companies on innovations;
–   identify and unlock new knowledge fields for companies.

5.3.3   Implementation of ETP

5.3.3.1   Implementation at the national level
Two organisations are mainly responsible for the execution of technology programmes at the national level. Senter is the administrative agency responsible for most programmes established by the Ministry of Economic Affairs, and has around 100 subsidy programmes, mainly focused on energy, and a annual budget of around €250 million. Novem executes environmental programmes established by the Ministry of Environment and is also involved in several energy programmes under the Ministry of Economic Affairs. Very large programmes, such as EET, have their own executive agencies, made up of professionals seconded from both Novem and Senter.

Both Novem and Senter act as spiders in the web of the various subsidy programmes for environment-oriented technology. The professionals in these organisations have to ensure the correct application of the procedures for the different schemes. Schemes often involve an application period of around three months, and then an advisory committee, or 'a board of wise persons', prioritise the different proposals using criteria set in the programme objectives. Alongside their work in guiding this process, and administrating incoming, ongoing and finalised projects, Novem and Senter also play an important networking role in the whole process of starting up technology projects. They act as intermediaries and offer advice to organisations that are considering setting up specific projects, and to companies which have specific questions regarding solutions for environmental problems, or technological demands. With their knowledge of most organisations involved in technology development, and because specific officers are specialised in certain technologies, they are often well placed to bring relevant parties into contact with each other.

5.3.3.2   Implementation at the regional level

In order to reach small and medium sized companies, technology policy also pays attention to the regional level. Therefore, both for technology and environmental technology, several activities are carried out at the regional level in the Netherlands.

We can crudely divide these activities into three main categories:

- specific tasks performed at the provincial level in the areas of cleaner production and pollution prevention;
- the establishment of agencies to facilitate the introduction of innovations in SMEs: the previous innovation centres, now called Syntens, and an innovation network for entrepreneurs;
- subsidy schemes for innovation, technology, and the environment at the regional level; with a specific focus on the structure and problems of the regional industrial and agricultural sectors.

The Dutch provinces carry out various activities related to the innovative behaviour of companies. The development in the 1980s of pollution prevention as a tool for improving environmental performance in companies has led to the setting up of pollution prevention teams in most provinces in the Netherlands. These teams initiate and stimulate various pollution prevention projects throughout the Netherlands (De Bruijn and Hofman, 1998). Various subsidy schemes are in operation in the provinces, with the majority of subsidies targeted at the economically less developed areas. Most of these subsidies focus on economic and technological development but, in every province, at least one scheme has a specific focus on the environment, innovation, or the prevention of pollution. The Departments of Economic Affairs and of Environmental Affairs, together with the consultation group of the provinces (IPO), engage in promoting projects for cleaner production. Initiatives towards cleaner production are supported both at the national and regional levels. Further, organisations such as Syntens, which operate regionally, play a role in these activities.

5.3.4   New developments in environment-oriented technology policy

Linkages between environmental policy and technology policy: branch covenants and related technology programmes. Negotiated agreements for industrial branches are among the recent innovations in Dutch environmental policy. Related to these covenants, through a specific technology programme, technologies are stimulated which have the potential to solve some of the bottlenecks for reaching the targets of specific branches. In exchange for the willingness of branches to commit to specific targets set down in the agreements, the government is prepared to support branches in the development of environment-oriented technologies. The technology programme provides an indication to the branch and to related technology developers the problems that need to be solved in the medium and long term.

A focus on technological breakthroughs through long-term commitments and with co-operation as a crucial factor: the EET and DTO programmes. In several programmes and policies, the EP and ETP time horizons increasingly include the longer term. In this longer term, efficiency gains by factors of 4 to 10 are aimed for. Therefore, in a programme such as EET (Economy, Ecology, and Technology, see also Table 5.4) the focus is on technological breakthroughs and long-term projects with the co-operation and commitment of different parties as critical conditions. This co-operation should encompass parties involved in fundamental research, such as universities; partners involved in strategic research, such as research institutes; partners involved in applied research, such as engineering firms; and finally companies which are marketing or using the technology. The time horizon of the DTO programme (Sustainable Technology Development) is even longer, from 20 to 50 years. This programme establishes scenarios in which efficiency improvements by factors of 10 to 20 are starting points. This programme tries to outline possible future lines of development, it does not develop the actual technology.

Cluster policy. This involves the stimulation of co-operation and network-forming between companies and knowledge institutes, and between various companies. Various policy instruments play a role in the stimulation of technological co-operation and the formation of innovative clusters. Lately, government policy in this area has intensified, with technology programmes where co-operation between companies and research institutes is a condition, and programmes which focus on the development of sustainable industrial estates. Furthermore, innovation centres (Syntens), institutes for SMEs, and regional development corporations play an active role in strategic cluster formation in specific regions. Three major roles of the government can be distinguished:

- creating the right conditions: improving market mechanisms and innovative potential through deregulation, generic technology programmes, and good infrastructure;
- operating as an intermediary between supply and demand: providing strategic information, facilitating platforms for parties to meet each other, initiating specific projects focussed on technological co-operation;
- acting as a demanding buyer: infrastructure projects, regulation, and tender procedures. This is based on the historical insight that economic and technological renewal often occurs through a customer with far-reaching demands.

Benchmarking as a policy tool to stimulate cleaner production and the development and application of new technologies. Benchmarking has recently been introduced by the government as a policy instrument to induce companies to produce more efficiently. A benchmarking study can provide information on the status of the national emissions per unit of production in a certain industrial sector or company, as compared to other countries. An example is to be found in an agreement between the Dutch government and industry to reduce CO_2 emissions. According to the 'Environment and Economy' policy document, Dutch industry is ambitious and wants to belong to the best in the world in terms of energy efficiency. The national government very much welcomes this from the perspective of environmental protection and the perspective of international competition through cost-savings. The general idea is that benchmarking could be an extra impulse for energy saving, by assessing energy performances abroad and comparing them with Dutch industry. The energy-intensive industries will work on a protocol for benchmarking with the national government. An independent institute will analyse how much energy Dutch companies use per product unit and compare this performance with other countries. If the Dutch companies are not leading in terms of energy efficiency, then additional measures have to be taken to catch up within a reasonable time limit. As a reward for fulfilling the promises made in negotiated agreements with the government, companies will not be confronted with any additional measures aimed at energy savings and CO_2 reduction.

5.3.5   Evaluation of environment-oriented technology policy in the Netherlands

Dutch environment-oriented technology policy contributes to bridging the gap between demand for, and supply of, environmental technology. The main instrument has been to subsidise promising environment-oriented technologies. However, the focus of this instrument has changed over time. Its focus is now prevention-oriented rather than the previous emission-oriented one. Closing the cycles of water and material uses in processes are now prevailing concepts; concepts that have been extended from process to products, and from individual companies to product chains and groups of companies. The rationale and motivation for this stems from two basic concepts: in the first place it is accepted that sustainability, and solving Dutch environmental problems, needs these kinds of changes; and secondly there is now a widespread belief that these changes enhance competitiveness in the long run, and create new market opportunities. In short, it is expected that this will create win-win situations.

Apart from financial support for the development of environment-oriented technology policy, several other policy instruments are utilised. Most of them focus on the provision of information and advice to organisations potentially, or effectively, involved in the development of environment-oriented technology. Examples include the specific centres such as Syntens and Infomil which, at the regional level, inform organisations about innovative possibilities, anticipated regulations, or about bodies involved in the development of environment-oriented technology. At the national level, Novem and Senter have similar functions.

5.3.5.1 Strengths and weaknesses of the Dutch ETP
An important question is whether Dutch environment-oriented technology policy can be considered as successful and effective. It is difficult to give a straightforward answer to this question. Various evaluations have taken place, although more are required, which have looked at the effects and effectiveness of different technology programmes (Arentsen, Bosveld, and Bressers, 1992; Arentsen and Hofman, 1996; Berenschot, 1995; RIZA, 1995; TNO, 1996; Van Seggelen, 1993; Willems and van den Wildenberg, 1993). All these evaluations come to a similar conclusion: it is difficult to precisely quantify the effects of a technology programme. An apparent deficiency in most technology programmes is the lack of an adequate monitoring system, especially with respect to the follow-up to sponsored projects. What can be said is that the effects vary among the different beneficiaries and also that different aspects need to be taken into account. Considering the beneficiaries for example, large multinational companies are investing significantly in new technologies, and their behaviour in this respect is little influenced by the availability of subsidies. For SMEs, however, subsidy schemes are often crucial for the feasibility of projects, but such companies are often scared off by the administrative burden of applying for subsidies (Van Seggelen, 1993; Arentsen and Hofman, 1996). For research institutes, the funding of projects through technology programmes can amount to a very substantial part of the total investment in the development of technology. Especially for the more fundamental and risky projects, support from government plays a very

important role. An aspect which needs to be taken into account, apart from the apparent environmental impact of the technology (through the specific development and application of the technology, and through the possible diffusion of this technology), is the way in which subsidies can influence the behaviour of the target group. The following aspects can be considered as relevant (Arentsen, Bosveld, Bressers, 1992; Arentsen and Hofman, 1996):

- making the development of environment-oriented technology possible by providing the necessary financial means;
- making parties aware of the possibilities in the development of ET;
- improving the cost-benefit ratio of the development of ET;
- improving the attitude of companies towards environmental problems and environmental regulation;
- improving the climate for negotiations between government and industry, and decreasing resistance to the application of other policy instruments;
- improving the contacts between the various organisations involved in a specific policy and technology field.

In practice, different technology programmes and ETP stimuli will derive strengths from several of the aforementioned aspects. A clear weakness of the ETP is the difficulty in relating the effects to the costs of the programmes. In other words, are the effects sometimes not insignificant compared to the overall costs involved in setting up and maintaining these programmes? Some of the other apparent weaknesses of the ETP concern the co-ordination between the various government agencies and among different technology programmes. Although Novem and Senter are the main organisations involved in the implementation of the different programmes, and this is a strong point of the Dutch ETP, the large number of programmes sometimes confuses the individual company looking for possible support. Coupled with the bureaucratic nature of some of the schemes, this explains the success of consultancy agencies that specialise in the application and execution of subsidy schemes.

   A more serious weakness is that the co-ordination among programmes and government agencies is limited. One result is that technology which is developed in the more fundamentally-oriented technology programmes, but which is not yet ready to enter the market, does not necessarily progress to programmes that focus on making technologies ready to enter the market. Promising projects are not automatically guided by different programmes, and this leads to unnecessary inflicted costs if such projects do not succeed in surviving without financial assistance. Another problem is the fact that often the barriers to the development of an environment-oriented technology are not so much financial but of another nature. An example is the development of more efficient wind turbines, a programme which is heavily subsidised by the Dutch government, where the difficulty in entering the market is due to problems in obtaining permits to erect wind turbines. Similar regulatory constraints can occur when companies develop technologies to process waste product streams into useful material, but are not allowed to transport these waste products because this requires specific permits which are not granted.

5.4   Co-ordination of environmental policy and environment-oriented technology policy

This section outlines some of the efforts made to co-ordinate and integrate EP and ETP. The most important feature of EP/ETP co-ordination is the four-year national environmental policy plans which are the responsibility of several Dutch ministries, under the overall responsibility of the Department of Environment. However, the integration of environmental concerns in policy fields other than the environment is still in an infant stage. As an illustration, recent research has shown that many of the intended effects of environmental policy instruments, such as to reduce household consumption and car mobility, have been offset by the unintentional side effects of other public policies (Ligteringen, 1999). This emphasises that efforts have to be made to explain the effects of policies, and to co-ordinate the use of different policies and policy instruments. Some examples of where steps have been taken towards co-ordination and integration are outlined below.

5.4.1   Integration of economic and environmental policy: the environment and economics policy document

EP/ETP co-ordination has become more prominent over the last decade, especially since EP and ETP are increasingly viewed as not only being potentially detrimental to economic growth but also as offering new perspectives and opportunities for more sustained growth and increased competitiveness. A further factor that is helping co-ordination between EP and ETP, and in particular between the Ministries of Environment and of Economic Affairs, is the climate change challenge. The Ministry of Environment is primarily involved in setting goals for CO_2 reduction, and the Ministry of Economic Affairs is responsible for energy policy, which makes co-ordination imperative. However, the most striking example of co-ordination has been the joint effort by the Ministries of Environment, of Economic Affairs, of Transport, and of Agriculture, to provide a perspective on the types of changes and renewals where economic growth, competitiveness, increased efficiency, and reduced environmental impact can go hand-in-hand. This has been laid down in the 'Environment and

Economy' policy document (VROM et al., 1997). Apart from having civil servants from the four departments working together to create different perspectives, this has also led to a series of specific policy initiatives. One overall principle that has been agreed upon is that taxes will have to shift from placing a burden on labour to laying the burden on environmentally unfriendly activities. The document also provides inputs for important infrastructure decisions that will have to be taken in the coming years.

Two comments that are more critical with regard to this document should be noted. Although it is a significant step and development towards removing the barriers between different departments and policy fields, it is clear that when it comes to actual decision-making that long-term environmental values are still secondary to the shorter-term economic concerns. Several decision-making processes on large infrastructure projects provide evidence of this. Secondly, although the perspective adopted in the policy document is very attractive, namely that both the economy and the environment can win, in real life this is often not the case. Further, there is a danger that strategies for economic expansion are being legitimised using the argument that they are also positive for the environment whereas, in practice, the environmental aspects of the expansion are not given adequate consideration.

5.4.2   Combinations of environmental and technology policy instruments: focus on specific target groups

An example of a co-ordinated effort to integrate EP and ETP is the negotiated agreements in industry and related technology programmes. Figure 5.2 provides an overview of the linkages between negotiated agreements, technology programmes, and direct regulation (the licensing scheme). Covenants have clear targets for specific themes and substances that are based on the overall targets in the National Environmental Policy Plan. Studies are carried out to assess the potential of specific branches to meet these targets. This calculation is based on the implementation and diffusion of the best technical means available throughout the branch. If this does not allow the targets to be reached, this indicates that certain technological bottlenecks exist that prevent the targets being met. Different programmes are then developed to stimulate the development of technologies that have the potential to remove these bottlenecks. Initially, a specific technology programme is set up to develop technologies that can overcome bottlenecks (Programme Environment and Technology, see also Table 5.4). Secondly, task forces are set up to inform industries of the scope for technological improvement, and a specific technology programme is established to stimulate initial applications of newly developed technologies

(Programme 'Referentieprojecten Milieutechnologie' by the Ministry of Economic Affairs).

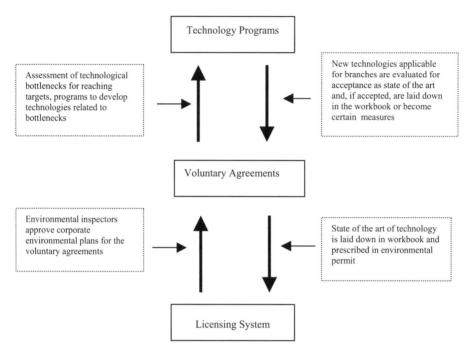

Figure 5.2: EP and ETP interlinkages in the voluntary agreements for branches of industry

Thirdly, for new technologies that are shown to offer a significant improvement in environmental efficiency, or to reduce environmental impact, a specific tax scheme has been developed. Such technologies are placed on an approved list, and companies then using them can write off their investment costs in the year that suits them best, thus offering them the potential of considerable tax savings.

Apart from the actual stimulation of technologies, the programmes also create a medium- and long-term perspective on the required technological changes and environmental problems that need to be solved. The downside is, however, that direct regulation and the setting of specific standards still play important roles in pollution control policies and can constrain innovative activities by companies.

5.4.3 Stimulating responsiveness and innovativeness of companies through environmental policy measures

A crucial element of a successful technology policy is that the companies who are expected to apply the technologies have the capability and willingness to innovate and implement organisational and technological changes, and modifications to their products, processes, and management. Several policies have been developed in the Netherlands which have the potential to foster these innovative capacities. The stimulation of the introduction of environmental management systems into companies, and of the use of the concept of pollution prevention in companies, has increased the sensitivity of companies to technological opportunities which have both environmental and economical benefits. In this sense, it can be argued that such EP policies lay the foundations for a more successful ETP.

5.4.4 Organisations at the interface between environmental policy and technology policy

Apart from the relatively direct stimulation of technologies, policies are also in place to inform companies of the options available for technological and organisational improvements, for example by providing examples of companies that have successfully introduced organisational and technological changes. Therefore, specific organisations play an important role in distributing these kinds of messages to companies. Materials such as books with success stories of companies have been developed (examples of active environmental approaches in Dutch business) (EZ and VROM, 1995). Specific organisations, such as Infomil and Syntens, at the interface between environmental policy and environment-oriented technology policy have been set up and stimulated. Dutch EP also makes use of the networks in which companies operate. EMS stimulation has made use of intermediaries such as branch organisations. Since messages conveyed by government agencies are often viewed suspiciously, companies are much more likely to accept a message and take appropriate actions if it is received from an organisation with which they feel much more at ease, and which they expect to represent their interests.

REFERENCES

Arentsen, M.J, Bosveld, D.J. and Bressers, J.Th.A (1992) *Een plaats voor de Stimuleringsregeling Milieutechnologie?* Enschede: University of Twente, Center for Clean Technology and Environmental Policy.

Arentsen M.J. and P.S. Hofman (1996) *Technologie, Schone motor van de economie?* Publicatiereeks milieustrategie, 1996/16. Den Haag: Ministerie van Volkshuisvesting, Ruimtelijke Ordening en Milieubeheer.

Berenschot, *Evaluatie en vervolg inzake het IOP-MT,* juni 1995.

Bressers, J.Th.A. (1983) *Beleidseffektiviteit en waterkwaliteitsbeleid.* Enschede: Ph.D. Thesis University of Twente.

Bressers, J.Th.A. (1988) A comparison of the effectiveness of incentives and directives: The case of Dutch water policy. *Policy Studies Review,* 7 (3): 500-518.

Bressers, J.Th.A. (1993) Beleidsnetwerken en instrumentenkeuze. In: *Beleidswetenschap,* 7 (4): 309-330.

Bressers, J.Th.A., D. Huitema and S.M.M. Kuks (1994) Policy networks in Dutch water policy. In: J.Th.A. Bressers, L.J. O'Toole jr., and J. Richardson (eds.), Water policy networks, special issue *Environmental Politics,* 3 (4): 24-52.

Bressers, J.Th.A. and Plettenburg, L.A. (1997) The Netherlands. In: M. Jänicke and H. Weidner (eds.) *National environmental policies. A comparative study of capacity-building.* Berlin: Springer, pp. 109-131.

Bruijn, T.J.N.M. de, and P.S. Hofman (1998) Varieties of Pollution Prevention, a Path towards Sustainable Development? In: *Proceedings of the 7th Greening of Industry Conference,* Rome, 15-18 November 1998.

Biekart J.W. (1995) Environmental covenants between government and industry. A Dutch NGOs experience. *RECIEL,* 4 (2): 141-149.

Bruijn, T.J.N.M. de, Lulofs, K.R.D. (1995) Netwerken rond milieuzorg. In: Bressers et al. (eds.) *Milieumanagement.* Alphen aan de Rijn: Samson Tjeenk Willink.

Bruijn, T.J.N.M. de, Lulofs, K.R.D. (1996) *Bevordering van milieumanagement in organisaties. Een kwantitatief onderzoek naar beleidsvoering met doelbewust gebruik van beleidsnetwerken.* Ph.D Thesis. Enschede: Twente University Press.

Carson, R. (1962) *Silent spring.* Greenwich (Conn): Fawcett.

Coenen, F.H.J.M (1992) *The role of municipal environmental policy planning in environmental management.* paper IIAS World Congress, July 1992. Enschede: University of Twente, Center for Clean Technology and Environmental Policy.

Cramer, J. (1996) Verandering binnen het milieutechnologiebeleid. In: J. Eberg et al. (eds.), *Leren met beleid.* Amsterdam: Het Spinhuis, pp. 127-141.

Dijk, J.W.A. van, N. van Hulst (1998) Grondslagen van het technologiebeleid. *ESB,* 21-9-1988.

Drees, W. (1992) Strategiën voor het NMP-2. *Milieu. Tijdschrift voor milieukunde,* 7 (2).

EZ, OCW & LNV (1995) *Nota Kennis in Beweging* (Policy Document Knowledge in Action), Den Haag.

EZ and VROM (1991) *Nota Technologie en Milieu* (Policy Document on Technology and Environment). Tweede Kamer der Staten-Generaal, vergaderjaar 1990-1991, 22 085, nrs 1.-2. 's-Gravenhage: Sdu Uitgevers.

Hofman, P.S. (1997) Ketengerichte Milieutechnologie. In: K.R.D. Lulofs en G.J.I. Schrama, *Ketenbeheer.* Enschede: Twente University Press, pp. 71-87.

Hofman P.S, Schrama G.J.I. (1999) *Innovations in the Dutch environmental policy for the industry target group,* Paper for the eighth Greening of Industry Network conference, Chapel Hill, USA, 1999. Enschede: University of Twente, Center for Clean Technology and Environmental Policy.

Kemp, R.P.M. (1995) *Environmental Policy and Technical Change.* Maastricht: Universitaire Pers Maastricht (PhD Thesis University of Limburg).

Koning, M.E.L. de (1994) *In dienst van het milieu. Enkele memoires van oud directeur-generaal prof. ir. W.C. Reij.* Alphen aan den Rijn: Samsom H.D. Tjeenk Willink.

Le Blansch, C.G. (1996) *Milieuzorg in bedrijven. Overheidssturing in het perspectief van de verinnerlijkingsbeleidslijn.* PhD thesis University of Utrecht. Amsterdam, Thesis Publishers.

Ligteringen, J.J. (1999) *The feasibility of Dutch environmental policy instruments.* Ph.D. Thesis University of Twente. Enschede: Twente University Press.

Meadows, D. (1972) *The limits to growth.* New York: Universe Books.

Moel, I. de, Jong, G. de, Nooteboom, S., Molenaar, R. (1999) *Opzet en structuur van het milieubeleid.* Amersfoort: DHV Milieu en Infrastructuur.

Mol, A.P.J., et al. (1998) *Joint environmental policy-making. New interactive approaches in the EU and selected member states.* Final Report. (Publications in Environmental Sociology, 11) Wageningen: Agricultural University Wageningen.

Nentjes, A. en Wiersma, D. (1988) Innovation and Pollution Control. In: *International Journal of Social Economics*, 15: 51-71.

Novem (2002) *Jaarverslag Groen Beleggen.* Utrecht: Novem.

Oorschot, P. van (1998) Snelle groei van ISO 14001. *Milieumagazine*, 1998 (1-2).

Opschoor, J.B. (1995) Economische instrumenten in het Nederlandse milieubeleid. *Milieu. Tijdschrift voor milieukunde*, 10 (5): 255-262.

Opschoor J.B. and Vos H.B. (1989) *Economic instruments for environmental protection.* Paris: OECD.

RIVM (National Research Institute for Public Health and the Environment) (1988) *Zorgen voor Morgen, Nationale Milieuverkenning 1985-2010.* Alphen aan de Rijn: Samson H.D. Tjeenk Willink.

RIVM (National Research Institute for Public Health and the Environment) (1998), *Milieubalans 98* (Environmental Balance Sheet 98). Bilthoven: Rijksinstituut voor Volksgezondheid en Milieu.

RIZA (1995) *Evaluatie van watergerelateerde projecten in de stimuleringsregeling milieutechnologie.* Lelystad: RIZA.

TNO (1996) *IOP-Preventie, Verbetering van succesfactoren van IOP-Projecten.*

Schrama, G.J.I. (2002) Stimuleren van duurzame innovaties door het MKB. *Beleidswetenschap*, 16(2): 162-181.

Seggelen, W. van, et al. (1993) *Evaluatie PBTS-Milieutechnologie.* Bakkenist-Micon, oktober 1993.

Tweede Kamer (1979) *Innovatie*, Tweede Kamer der Staten-Generaal, vergaderjaar 1979-1980, 15 855, nr. 2. 's-Gravenhage: Sdu Uitgevers.

VROM (1989a) *Nationaal Milieubeleidsplan 1990-1994.* Kiezen of Verliezen (NEPP 1). Tweede Kamer der Staten-Generaal, vergaderjaar 1988-1989, 21 137, nr.2. '-Gravenhage: Sdu Uitgevers.

VROM (1989b) *Notitie bedrijfsinterne milieuzorg.* Tweede Kamer der Staten-Generaal, vergaderjaar 1989-1990, 20 633, nrs.2-3. 's-Gravenhage: Sdu Uitgevers.

VROM (1990) *Nationaal Milieubeleidsplan-plus* (Annex to NEPP 1). Tweede Kamer der Staten-Generaal, vergaderjaar 1989-1990, 21 137, nr.21. 's-Gravenhage: Sdu Uitgevers.

VROM (1993) Nationaal Milieubeleidsplan 2. Milieu als maatstaf (NEPP 2). Tweede Kamer der Staten-Generaal, vergaderjaar 1993-1994, 23 560, nr.2. 's-Gravenhage: Sdu Uitgevers.

VROM (1997) *Nota Milieu en Economie. Op weg naar een duurzame economie* (Policy Document on Environment and Economy). Tweede Kamer der Staten-Generaal, vergaderjaar 1996-1997, 25.405, nr. 2. 's-Gravenhage: Ministerie van Volkshuisvesting, Ruimtelijke Ordening en Milieubeheer

VROM (1998a) *Nationaal Milieubeleidsplan 3* (NEPP 3). Den Haag: Ministerie van Volkshuisvesting, Ruimtelijke Ordening en Milieubeheer.

VROM (1998b) *Handboek Milieusubsidies.* Den Haag: SDU Uitgeverij.

VROM (1999) *Milieuprogramma 2000-2003*. Tweede Kamer der Staten-Generaal,
  vergaderjaar 1998-1999, 26 804, nr.2. 's-Gravenhage: Sdu Uitgevers.
VROM (2001) Nationaal Milieubeleidsplan 4. Een wereld en een wil. Werken aan
  duurzaamheid. (NEPP 4). Den Haag: Ministerie van Volkshuisvesting, Ruimtelijke
  Ordening en Milieubeheer.
Weaver, P., Jansen, L., Grootveld, G. van, Spiegel, E. van, Vergragt, P. (2000) *Sustainable
  technology development*. Sheffield: Greenleaf Publishing.
Winsemius, P. (1986) *Gast in eigen huis. Beschouwingen over milieumanagement*. Alphen
  aan den Rijn: Samsom H.D. Tjeenk Willink.
Willems en Van den Wildenberg BV (1993) *Evaluatie IOP-Milieubiotechnologie*.

Chapter 6

Environmental Policy and Environment-oriented Technology Policy in Spain

JOSE CARLOS CUERDA, MARIA JOSE FERNANDEZ
Institute for Regional Development, Fundación Universitaria, Spain

JUAN LARRAÑETA
Escuela Superior de Ingenieros de Sevilla, Spain

SUSANA MUÑOZ, FLORENCIO SANCHEZ, CARMEN VELEZ
Institute for Regional Development, Fundación Universitaria, Spain

6.1   Introduction

This chapter briefly presents Spanish environmental and technology policy with a special emphasis on issues relating to private businesses and the links between both policy systems.

We focus on the objectives, instruments, and actors involved in environmental and technology policy and the decisions made by the Spanish government. The reader should not forget that the decision making process in Spain is somewhat peculiar: there are 17 Autonomous Communities (CC.AA.), which have the authority both to design and implement the different policies. The rate at which power has been transferred from the national government to the Autonomous Communities has been different in each case and the process is still not completed yet. However, the Spanish Constitution confers on central government exclusive powers in certain areas. Several committees have been set up between the central government and the Autonomous Communities in order to adopt joint decisions on each

163

Geerten J.I. Schrama and Sabine Sedlacek (eds.) Environmental and Technology Policy in Europe.
Technological innovation and policy integration, 163-196. © 2003 Kluwer Academic Publishers. Printed in the Netherlands.

subject. Unfortunately, many conflicts arise during the decision making process, which therefore ends up being slow and complex.

This chapter is structured in four main areas: environmental policy, technology policies, the interactions between them, and some preliminary conclusions. The first two parts present general issues, the main challenges, the actors involved, objectives and instruments, and the interaction between Spanish and EU policies.

The risks associated with any R&D activity and the application of its results form one of the arguments for public intervention, due to the existing market inefficiencies in this type of activity. Thus, one of the objectives governments must have in R&D issues is to promote those activities that the market does not undertake by itself.

On the other hand, technological progress is a critical element in the long-term conservation of resources and the environment. Technology alone is not an autonomous force since it reflects the economic, social, and organisational priorities of industrial societies. Therefore, if it is adequately managed with the appropriate incentives and analyses, technology can help to solve the environmental problems it has been accused of generating. The scope of options that can play a role depend to a great extend on success in determining the cause-effect relations and in searching for adequate scientific and technological solutions. The determination of these options and the analysis of their effectiveness is one of the objectives of the ENVINNO Project.

6.2   Environmental Policy

This section analyses the environmental policy implemented in Spain in recent years. After a brief presentation of the main orientations of the government's environmental strategy, we discuss the role of the different agents involved in the decision-making process – basically public institutions and business sector. The section on objectives and instruments discusses the objectives of the different environmental policies in order to meet the challenges that have been identified. Regarding instruments, we differentiate between command-and-control instruments, and market instruments, with a special emphasis on new developments.

Our focus is on the environmental policy and the decisions made by the Spanish central government. We concentrate on the actions promoted by the Ministry of Environment, the main institution responsible for Spain's environmental policy: other agents that operate within this policy system will be also discussed, however.

6.2.1   Policy systems

6.2.1.1   General environmental issues

In the Sectoral Conference on the Environment held in November 1994, the Secretary of State of Environment and Housing of the Ministry of Public Works, Transport and Environment (which was the ministry with decision making powers at that time) presented the basic guidelines of the National Environmental Strategy of the Spanish government by which it was to fulfil the commitments established by the Constitution[1] and in the international context. This strategy defined the following general objective: "*to guarantee the constitutional mandates and compliance with international agreements, promoting the achievement of sustainable development by means of a rational use of natural resources, environmental protection and creation of jobs linked to environmental priorities in all areas of society.*"

In general terms we can state that the environmental policy implemented by the Spanish government has changed its orientation over the last few years. Originally it focused on pollution control based on the application of end-of-pipe technologies which merely transformed pollution and made it look different; it then changed towards a new paradigm that promotes a better use of resources and focuses on the production of cleaner technologies that enable a reduction of pollutants' emissions. There has been change from an almost exclusive role of command-and-control environmental instruments to a wider use of market instruments, based essentially on entering into voluntary agreements between the business sector and the government. Despite these orientations and political challenges there is still a long way to go, since the effective adoption of these new trends by companies is not yet widespread.

6.2.1.2   Sectoral issues

The main environmental challenges in Spain regarding the industrial issue are those related to water quality and the management of water resources, waste management, air quality, noise prevention and reduction, energy saving and efficiency and use of renewable energies. Therefore, the priority action fields for the national environmental strategy are:

Water quality and management and use of water resources: In order to achieve the protection of water quality and the aqueous environment, the

---

[1]   Under the title Ruling Principles of Social and Economic Policy, the Spanish Constitution includes the environment, defining it as a public good that everyone is entitled to enjoy and is obligated to protect and preserve, and public powers must therefore watch over its rational use.

objectives and intervention guidelines of the Spanish government focus on a progressive reduction of pollution load by means of a greater control of water quality and appropriate treatment of sewage water and sludge.

One of the main actions of the government has been the construction of large infrastructural works (dams and transfer channels) in order to increase the quantity of available water; however, this measure has not definitely resolved any water scarcity problems. On the other hand, at local level actions have been started in order to meet the objectives proposed by the European Community Directive 91/271 on the sanitation and treatment of sewage water from urban settlements and industries.

Waste management: The commitments established by community directives on packaging and on waste disposal dumps and the current regulation on toxic and hazardous waste are the main references within which the Spanish government seeks to define its strategy and lines of action this field. This strategy is based on the application of preventive policies aimed at reducing the volume and toxicity of waste following the principle that reduction is the best possible management method.

Furthermore, the national and regional governments are taking the most convenient measures in order to reduce and prevent waste production at source. The purpose is to reduce the volume of waste generated, emphasising the need for shared responsibility and adequate environmental treatment.

Air quality and noise reduction and prevention: One of the main challenges that environmental policy faces in this field is the absence of information of emissions. This is the case especially for urban areas where, as a result of traffic and industrial emissions, the creation of an index of pollution sources is absolutely essential. Additionally, one of the lines currently followed is the establishment of an emission monitoring and control network and the provision of measurement equipment.

The treatment of industrial air pollution is one of the priority subjects of the national environmental strategy. The general objective is to conduct an inspection of the industrial sector and watch over its compliance with the current regulations on pollutant emission to the atmosphere.

Energy saving and efficiency and use of renewable forms of energy: One of the basic pillars of Spanish environmental policy is the use of renewable forms of energy, together with energy saving and efficiency. These are domestic resources, the use of which facilitates an increase degree of energy self-supply and independence, which would help to raise the competitiveness of Spanish companies in international markets. Nevertheless, there are obstacles caused by the lack of technical harmonisation as a result of the differences among national rules and regulations, or even their total absence.

In this field the main challenges are, first, to encourage energy saving and substitution of the different energy resources by giving incentives to options that are more efficient but suffer from difficulties in entering the market; and second, to harmonise regulations in this field.

6.2.2   Actors and processes

6.2.2.1   Actors
Public institutions are in charge of establishing objectives, instruments and working guidelines. Thus, the role of the government is an essential one, not only from the legislative point of view, but also in planning and eliminating obstacles in order to enable private companies to perform their role.

Decision-making powers are widely scattered: at national scale the institution responsible for regulation is the Ministry of Environment established in 1996,[2] which shares sectoral responsibilities with the Ministry of Agriculture and the Ministry of Industry and Energy. Other ministries whose actions affect environmental policy in a more collateral way are the Ministry of Education and Science, the Ministry of Public Works and Transport, and the Ministry of Economy and Treasury.

The Ministry of Environment is made up of:
– A Secretariat of State of Water and Coasts, whose tasks are the planning of water resources, construction of central government-dependent water infrastructure, the drafting and application of regulations on water and coasts, and co-ordination with the Autonomous Communities in the areas of sanitation and water treatment.
– A Secretariat General of Environment. This is the principal body of the Ministry of Environment and directs and co-ordinates the implementation of policy powers in this ministry.
– Vice-Secretariat of the ministry. Its tasks include the representation of the ministry, and the direction and supervision of the management centres as well as of the units reporting directly depending to it.

There are other institutions that also depend on the Ministry of Environment, such as the National Weather Institute, the National Water Commission and the Environment Advisory Council.

---

[2]   Before its creation, the Ministry of Public Works, Transport and Environment (MOPTMA) was responsible for environmental issues from 1993.

The *Sectoral Conference for Environment* is in charge of co-ordinating actions among the different regional governments. It is made up of the general directors of environment of all Autonomous Communities In spite of this co-ordination, many conflicts arise in the decision-making process, which creates a slow, complex process, rife with conflict.

The *Sociological Research Centre* (CIS). Its purpose is to study Spanish society, mainly by means of survey research studies. One of the main topics dealt with is concern for the environment and for technological innovation. This resulted in the timely disclosure of barometers and the creation of databases. The CIS is leading several programmes aimed at encouraging and promoting research in the social sciences. Results are disseminated to national and foreign scientific associations. The information collected and processed on environmental issues is sent to EUROSTAT.

The *Environment Advisory Council* (CAMA) is in charge of promoting citizen participation in the decision-making process on environmental issues. However, from its inception it has been bitterly criticised by the main Spanish environmental organisations.

6.2.2.2   *Decision making process*
The Ministry of Environment is empowered to draft basic environmental legislation. Central government is also responsible for transposing European Union directives into Spanish law. The Autonomous Communities are empowered to develop this legislation in greater depth; they can draft additional protective measures and are in charge of managing all those questions for which local entities are not responsible. Moreover, each Autonomous Communities can draft and implement its own environmental policy, which must be in conformity with the national one. Local administrations are responsible for implementing many environmental projects, but they frequently lack the financial capacity to do so. Therefore, the national, regional and local administrations are responsible for the basic guidelines of the environmental policy, defining the general framework for priority actions.

The task of the *Secretariat of State of Environment* of the Ministry of Environment is to co-ordinate the administrations with environmental policy powers and among these and the rest of the administrations affected. The body in charge of performing those co-ordination tasks as the backbone of the environmental policy is *the Sectoral Conference on Environment*.

The general idea is that the drafting of environmental policies in Spain should be based on the debate between the government, at different levels, and the different stakeholders. This is the purpose for which the

Environment Advisory Council (CAMA) was created. The Environmental Advisory Council includes experts, environmentalists, Non-Governmental Organisations (NGOs), consumer organisations, etc., together with representatives of all ministries with competence on environmental issues. This Council was established in July 1994 to act as the co-ordinating mechanism for drafting and monitoring environmental policy and proposing measures aimed at improving compliance with the international organisations regarding sustainable development. The problem is that this is merely an advisory body and its decisions are not binding on the Ministry of Environment.

6.2.3   Objectives and instruments

This section describes to the fundamental objectives pursued by environmental policy in Spain. We review the main sectoral planning instruments established by the Spanish Government. Next we present the main instruments – both command-and-control and market-oriented ones – used in Spain. The first group of instruments has been and still is the most frequently used type in Spain. However, over the last few years we have been able to perceive a new trend towards a wider use of instruments that could be labelled as voluntary ones, in which the key elements are agreements between companies and government, and integrated pollution control.

6.2.3.1   *Objectives and planning instruments*
Objectives: The philosophy underlying Spanish environmental policy focuses on raising the citizens' quality of life, supporting private initiatives and encouraging the whole of society to play a role in environmental policy by means of information, environmental education and dialogue with all interested social sectors and NGOs with a shared responsibility.

The objectives of the Ministry of Environment are to preserve and improve environmental quality, to contribute to human health and to guarantee a rational and efficient use of natural resources.

Planning instrument: Environmental policy in Spain is articulated around the different priority environmental action areas:

As far as water is concerned, the *water* policy rests on two basic pillars: a conservation policy (optimising existing infrastructures in order to obtain maximum performance and to raise public awareness regarding the savings policy) and a quality policy included in the National Wastewater Sanitation and Purification Plan (1995–2000).

Other basic planning instruments for water are the *National Water Plan* and *the Basin Water Plans*. The approval of these Plans was intended to introduce some degree of rationality in the construction of water public works, aside from laying down a set of prescriptive actions for the prevention of floods, spates, droughts and other types of catastrophe. This is a preparatory step to the approval of the National Water Plan, which will become operational in the near future.

Regarding *waste*, the main objectives pursued are the reduction of waste production and toxicity and a greater degree of waste control and management. For this purpose the *National Industrial Waste Plan* (1989) and after that the *National Plan for Toxic Waste* (1995–2000) came into operation. At the same time, some Autonomous Communities drafted their own *Territorial Waste Management Steering Plans* making use of their policy powers in this area.

The recovery of polluted soils is closely linked to the management of hazardous waste. The first actions in this field came out of the Identification, Control and Recovery Programme for areas affected by toxic and hazardous waste (included in the National Industrial Waste Plan of 1989). To start up this programme the former Ministry of Public Works, Energy and Environment started to draft an Inventory of Polluted Soils in 1991. The *National Plan for the Recovery of Polluted Soils* (1995–2005) articulates the mechanisms required for undertaking the decontamination works in the main target areas identified by the inventory and its purpose is to prevent soil pollution and to decontaminate those that were already polluted.

In the field of *energy*, after the fuel price shock, the national government started the National Energy Plan in 1978 which was operative for three years. The next one was the National Energy Plan of 1983 which was not sufficiently effective to solve the energy problem. In 1991 the National Energy Plan for 1991–2000 was established. We should also mention the *Energy Saving and Efficiency Plan* (1991–2000). This is part of the National Energy Plan of 1991, which defines the strategy for the efficient use of energy and the use of renewable energies (Renewable Energy Programme). Thus, it is an incentive to extend alternative types of energy, to substitute the most polluting electric energy generation equipment and to reduce global emissions of volatile organic compounds, carbon monoxide and carbon dioxide and sulphur and nitrogen acids.[3]

---

[2] In this sense, it is essential to take into account community regulations, especially the IPPC directive, with a great impact on the industrial sector and especially on the energy sub-sector.

In the area of *air pollution* there are no national plans since it is the Autonomous Communities that transfer this policy power to the municipalities. However, air pollution is handled by different sectoral plans, and the main referent is the Law 38/1972 on Protection of the Air Environment.

In addition, Spain obtains important resources from the EU to finance environmental projects, basically from the Structural Fund and the Cohesion Funds of which Spain is the largest beneficiary (27% of the total). The total sum allocated to environmental projects in Spain from the Structural Funds for the period 1994–1999 accounts for €2,764.66 million out of a total sum of €0.03 billion.

6.2.3.2   Command and- control oriented instruments

When speaking about command and control instruments we refer to the traditional regulation of environmental issues. Among the main problems that command-and-control instruments have are the difficulty in controlling compliance, the high cost of their implementation, and the enormous bureaucracy they cause, with a resulting adverse effect on companies.

Spain has very numerous, scattered items of environmental legislation because there are many institutions that have or hold regulatory powers on environmental issues, and these regulations are scattered throughout laws, regulations or legal texts, the main focus of which is on other issues than environmental protection.

The most relevant environmental regulations for the purpose of this chapter are:

Water. The most important regulations on water are the Law of Water 29/1985, of August 2nd [4] and the Executive Order 849/1986 which approved the Public Water Domain that develops the Law. The objectives pursued by this law are a rational use of water, an adequate protection of the resource and to ensure the availability and quality of water without degradation of the environment. The law defines the tasks of the state and lists, *inter alia*, hydrological planning, drafting of the water infrastructure plans and compliance with international agreements and covenants on water. Each Autonomous Community has policy powers on water, which are different according to their statutes.

Executive Order 849/1986, Regulation of the Public Water Domain, which develops the Law of Water, establishes quantitative and qualitative limits for sewage water as well as the requirement for the prior authorisation

---

[4]   The draft for the reform of this Law was presented in May 1997.

of the operation of an industry or activity that is established, modified or moved.

It is also necessary to mention Executive Order 11/95, which transposes European Community Directive 91/271/EEC on urban sewage water treatment in order to protect the quality of continental and sea waters from the negative effects of urban sewage waters.

Waste. The Waste Law 10/1998 of April 21st applies to all types of waste except air emissions, radioactive waste, and sewage water. This law repeals all previous laws that separately treated urban solid waste and toxic and hazardous waste.[5] It does not just regulate wastes after generation; it also considers them in the stage prior to their generation. It regulates activities of producers, importers, and intra-community purchasers, and in general, of any person bringing to the market products that generate waste. The law also foresees the treatment of polluted soils. Thus, it establishes that the Autonomous Communities must draft a list of action priorities considering the risk soil pollution entails for human health and for the environment.

The Law 11/1997 on Packaging and Packaging Waste was issued on April 24th, and the Regulation that develops it was approved by Executive Order 782/1998 of April 30th. This law is the result of the transposition of the packaging (wrapping, bottling, tinning, canning) and packaging waste law of the EU directive into Spanish legislation. Its main objectives are to protect or to reduce the impact of these wastes on the environment, following the principle that reduction is the best possible means of waste management, as assumed by the Community Strategy on waste and the 5th Community Action Plan on environment and sustainable development. The law establishes the obligation that packaging should be manufactures without exceeding certain levels of heavy metal concentration and following certain technical requirements. As a complement to these preventive measures, the regulation promotes reuse and recycling of used packaging as the best available reduction method. For this purpose Public Administrations can establish economic, financial and tax measures.

Air pollution. The regulation on air pollution (Law 38/1972 of December 22nd, of Air Environment Protection and other specific technical regulations) establishes air quality standards and limitations to certain emissions and specifies the character and tasks of the national follow-up and quality control network. The law enables the government to award subsidies and other financial aid to facilitate to companies the compliance with their legal obligations and there is also support for investments in equipment to control air pollution. Furthermore, tax reductions and soft loans can be awarded.

---

[5]   Law 42/1975 of urban solid waste and Law 20/1986 of toxic and hazardous waste.

However, environmental legislation changes very much from one region to the other, due to the peculiar features of each region and to the delegation of policy powers to municipalities, which are the bodies that implement the specific treatment. For example, in the Basque Country air emission standards are almost equivalent to those of Northern Europe, while in Catalonia the regulations are less stringent.

Energy. A recent action in the field of energy has been the implementation of measures aimed at introducing competition in the electric sector (Protocol for the establishment of a new regulation of the national electric system) and in the sectors of natural gas (Executive Order on the access of third parties to the national gas pipeline and gasification plant network) and of petroleum derived products (Executive Order on the access of third parties to oil product and liquefied petroleum gases).

The national administration, by means of the Ministry of Industry and Energy (MINER), encourages the development of renewable energies by means of the promotion of self-production of electric energy, via prices. For this purpose MINER issued an executive order on electric self-production and co-generation. This new regulation unifies in one single Executive Order all the basic regulations that give a specific treatment to each type of facility. The scheme proposed consolidates and increases in general terms the prices of the above means of production and facilitates the maximum development of the energy planning guidelines included in the current National Energy Plan (1991–2000).

These actions of MINER have been complemented by the promotion of the energy saving and efficiency strategy, the main objective of which is environmental protection by means of positive actions to encourage saving and substitution of energy resources that give incentives to the dissemination of more efficient technologies. The Global Subsidy ERDF-IDEA is managed with the R&D national and community promotion programmes in the energy sector for this purpose.

Another set of regulations includes environmental aspects in their articles. Among others, the Law 21/1992 of Industry, which provides the basic legal framework for industrial activities in Spain, one of its tasks being to contribute to make compatible industrial activities and environmental protection. The law previsions the drafting of plans to improve competitiveness and the adaptation of industrial activities to national and European Union regulations on environmental issues.

On the other hand, with the Organic Law 10/1995 of the Criminal Code becoming effective, there is a substantial change regarding the regulation of environmental offences. These are listed as follows: offences against territorial organisation and urbanism, offences against natural resources and the environment, offences against flora and fauna protection, offences regarding nuclear energy and ionising radiation, other risk offences and forest fire offences.

6.2.3.3   Market oriented instruments

Aids and Subsidies. The main feature of Spanish environmental policy is its financial dependence on European funds, especially through the Structural Funds (ESF, ERDF, and EAGFL-G) and the Cohesion Fund. There are EU economic instruments that can be directly accessed by Spanish companies. The most relevant ones in environmental issues are LIFE, ALTENER, SAVE and the 4th and 5th EU R&D Framework Programme.

Public subsidies to environmental investments are conditioned by EU regulations and are limited to 15% for large companies and 40% for small-sized companies. In Spain the Ministry of Industry and Energy has developed the ATYCA Initiative (Plan to Support industrial Technology, Safety and Quality), which includes a corporate support programme. But following the general European trend the use of this type of instruments will decrease progressively in favour of other instruments.

Duties, Charges, Fees and Taxes. The use of these instruments is very limited in Spain. They are used only in some Autonomous Communities and just 6% of the central government's expenditure in environment is financed through fees and taxes, and the primary task is to collect funds.

The main reasons for using this type of instrument are the need to internalise environmental externalities, to give incentives for environmental behaviour and especially to produce income for the administration. The Ministry of Economy and Treasury maintains that the funds collected by these instruments do not have to be used for specific purposes, including environmental ones, but they must go into the Public Treasury.

The effectiveness of duties and fees is quite limited, due both to the difficulty of managing them and to the minor importance of funds collected. The clearest example of the use of this instrument is the emission duty, the water fee and the collection charge for domestic waste or solid waste assimilated to urban waste.

Tax Allowances. The government issued the Executive Order 1594/1997 which regulated the deduction for investments aimed at environmental protection. It provides for a deduction of 10% of the total sum invested in tangible fixed assets for the purpose of correcting environmental impacts.

Although the tax incentive is justified by the benefit the community receives from the improvement of environmental management standards, the right to deduction, as the logical consequence of the 'polluter-pays principle', is conditional on the maintenance of the environmental protection levels previously established by the relevant Administration during the term of the investment's duration.

The investments subject to tax deductions are those allocated to environmental conservation that take the form of facilities with one of the following objectives: to prevent or reduce air pollution coming from industrial facilities; to prevent or reduce the pollution load discharged on surface, underground or sea water; and to favour the reduction, recovery or correct treatment of industrial waste from the environmental point of view.

Sectoral Agreements. Voluntary agreements are being increasingly used by public administrations to apply their environmental policies, especially as far as industries are concerned. The first agreements signed aimed at the progressive elimination of CFCs in aerosols and at waste management. They were signed by the Ministry of Public Works, Transport and Environment (currently the Ministry of Environment). Other more recent agreements aimed at the withdrawal of old automobiles (RENOVE Plan I and II) and old tyres. The central administration has set up different initiatives aimed at the establishment of sectoral agreements (electric and electronic material, paints, chlorinated products, etc.).

On the other hand, Executive Order RD 484/1995 acknowledges the needs of industry to adapt in the fields of sewage water treatment and control, authorising Water Confederations to negotiate and sign agreements with industrial associations to control sewage water.

Eco-labelling. This is a voluntary market instrument with a corrective character and an information purpose, which facilitates consumer information and selection capacity, and encourages producers and distributors to increase their market share, improving production processes and reducing any environmental impacts caused.

Executive Order 598/94 issued in Spain in 1994 established the rules for the application of the EU Regulation 880/92 on the community award system of the environmental/ecological label. Despite the expectations raised, there have been many difficulties in its application, resulting in a very small number of products that have obtained the label.

At the national scale, AENOR (Spanish Standards Association), in collaboration with MINER, has created a Spanish eco-labelling system under the brands AENOR-Environment. Catalonia is the only autonomous community that has developed its own ecological labelling system: 'Environmental quality distinction'.

Environmental Management System and Environmental Audit. The
environmental management systems (SGMA) are management tools used for
attaining the objectives established by the environmental policy. Currently,
the most important regulations are EEC Regulation 1836/93 (EMAS) (in
Spain, Executive Order R. D. 85/1996) and the standard ISO 14001/96
which has absorbed the standard UNE 77801/94, from then on being called
UNE-EN-ISO 14001.[6]

Although the introduction of the ISO 14001 standard started very slowly
in Spain, industry has become aware of the importance of this instrument
and the number of registered companies is currently growing. Regarding the
community management system EMAS, its degree of adoption is much
lower and most companies that install it are transnational corporations.

Regardless of the direct application of EU Regulation 1836/93 on the
community environmental audit and management system, the Spanish rule
UNE 77-802-93 establishes the general rules for environmental audits and
environmental management system audits. This is the national rule that deals
with environmental audit. At regional scale, the Autonomous Communities
have policy powers on legislative development in environmental audit
matters. However, very few of them have passed legislation on
environmental audits.

Environmental Impact Assessment. Environmental impact assessment is a
corporate environmental management instrument with a preventive character
and a managerial task, consisting of a process the purpose of which is the
identification, prediction and interpretation of environmental impacts of a
project or activity.

Executive Order 1302/1986 on environmental impact assessment
incorporates EEC directive (85/337/EEC) into Spanish Law. This regulation
orders the implementation of environmental impact assessment on public
and private projects consisting of the construction of public works, facilities
or any other activity included in the annex to the EU directive, and also in
large dams, first reafforestations under certain circumstances, and the open
air extraction of minerals.[7]

---

[6] The European management system (EMAS) is much more demanding than the ISO norm,
especially in aspects such as continuous improvement, pollution prevention and
environmental information. Due to the lower level of requirements, companies use to
introduce the ISO 14001/96 system.

[7] We should highlight the legislation that has been developed by most Autonomous
Community, improving national law on this field.

6.3   Science, technology and innovation environment-oriented policies

Spain has no long-term tradition in R&D policies. While more developed OECD countries established co-ordination and management policies in order to promote innovation, Spain did not undertake this task until the second half of the 1980s, especially after the approval of the Science Law of 1986. Since then the situation has evolved quite positively. Spain's entry into the EEC in 1986 and the concern shown by the Autonomous Communities have largely contributed to this evolution.

The same turning point finds reflection in the Spanish industrial and technological policies. As we shall see in this section, there is a perceived change of trend as far as the environmental policy is concerned. Despite the fact that the Industry Law of 1992 foresees the drafting of plans aimed at improving the competitiveness and adaptability of industrial activities to national and EU environmental regulations, the main reference can be found in the establishment in 1990 of the Environmental Industrial and Technological Programme (PITMA) for the 1990–1994 period with a follow-up for the 1995–1999 period. This programme was incorporated after the creation by the Ministry of Environment of the community initiative ATYCA (Plan to support Industrial Technology, Safety, and Technology) in 1997, which is also discussed in this chapter.

We shall deal with those policies promoted by Central Government, especially by the Ministry of Industry and Energy (MINER), the Ministry of Education (actions on the field of science, research, and technological development), and also by the Ministries of Economy and Environment. Other actions promoted by other ministries also affect industrial, technological, and scientific policies. Such is the case with the policies implemented, for example, by the Ministry of Labour.

6.3.1   Policy system

After covering the first stage, in which the Spanish scientific policy has focused on the promotion of basic research and the setting up of infrastructures, the Spanish government has recently outlined its intention to modify its policy orientation in order to place a greater emphasis on technological innovation: influencing the activities of research teams by directing them to the research areas that appear the most promising, and promoting corporate research within our borders. This new orientation has arisen after it was concluded that there are several deficiencies in the science–technology–industry system in Spain: powers on R&D issues are scattered over several government levels (ministries and Autonomous

Communities); there is little participation of companies, especially SMEs, in public R&D programmes; there are deficiencies in the transfer of results from basic research to the industrial fabric; and finally, a significant gap remains between Spain's R&D effort (both public and private) and that of the average developed countries.[8]

One of the main challenges industrial R&D policy faces is that companies assume a leading role in innovation by means of a set of supports and incentives. It is all about promoting a favourable environment for innovation, acting in those fields where private initiative has a minor presence. The main objective is to promote technological innovation, quality and security as the pillars for generating competitive advantages in Spanish companies.

The integration of the environmental dimension into the technological policy is a determinant element to reduce environmental impacts. On the other hand, the consideration of the environmental variable in the general business policy means a new challenge for companies, since it means a new framework for competition from the point of view both of supply and of demand. This new framework results from the globalisation of environmental costs; the need to meet the new consumer demands, increasingly concerned with the benefit of using more environmental friendly products; and a reorientation of production technologies towards the reduction of pollutant emissions and the saving of resources.

6.3.2   Actors and processes

6.3.2.1   Actors
The participation of the Spanish administration in the Spanish innovation system was first regulated in 1986 with the approval of the Science Law, which provided for the creation of the Inter-ministerial Commission of Science and Technology (CICYT), two Advisory Councils and the National R&D Plan, the Ministry of Education becoming the main responsible institution for scientific policy (Fundación COTEC, 1998a).

Regarding the participation of the administration in industrial policy, the Ministry of Industry and Energy (MINER) is the responsible body. The orientation of MINER in terms of technology and innovation policy for

---

[8] Nevertheless, research in Spain has experienced a strong development over the last few years. Staff working in R&D amounted in 1985 to 63,109 employees (40,654 FTE Full Time Equivalent), and in 1995 total staff had increased to 147,046 (79,987 FTE). As far as number of researches is concerned, figures have gone up from 40,848 to 100,000 (47,342 FTE) over the same period (CICYT, 1996). Total expenditure has grown in a similar proportion. Over the 1988–1992 period, annual growth rates of total expenditure on R&D have averaged some 10.1%. However, since 1993 total expenditures as a percentage of GDP have remained fairly constant (OECD, 1996).

industrial competitiveness is based on a more effective articulation of the science, technology, and industry system, in the dissemination and promotion of an innovation culture that promotes entrepreneurial attitudes in society, and the establishment of a legal, regulatory and financial framework to favour innovation.

Moreover, the central administration participates in the science–technology–industry system through other ministries such as Economy and Treasury. One of the clearest examples for the purpose of this chapter is the setting up of the Industrial Development SME Initiative.

On the other hand, autonomous administrations have included the promotion of science, technology, and innovation among their objectives. In general, autonomous innovation policies have focused on technological development in industry and have become established in terms of support to companies and the setting up of infrastructure to support innovation (Fundación COTEC, 1998a). However, the effort devoted to R&D in Spain is very unevenly distributed among the different Autonomous Communities, which indicates insufficient inter-territorial cohesion in this field. The Autonomous Community that takes the highest share of Spain's total effort is Madrid, followed by Catalonia, the Basque Country, and Andalucia. This uneven distribution in total expenses reflects the geography of public university and non-university R&D centres and of private R&D centres (COTEC).[9]

6.3.2.2   Decision making process

Institutions. The 'Law for Promotion and General Co-ordination of Scientific and Technical Research' sets up a series of institutions in order to co-ordinate, assess and direct research in Spain. Some of these institutions are:

– The Inter-ministry Science and Technology Commission (CICYT). This commission is in charge of establishing the broad objectives of research and co-ordinating this research and it is also in charge of drafting, co-ordinating and monitoring the PN I+D. The CICYT is chaired by the prime minister and drafts the general guidelines of scientific policy and it has an administrative support organisation, the Secretariat General of the PN I+D, which falls under the Ministry of Education.

---

[9]   However, these territorial imbalances have been improved in recent years, due mainly to: a) The extension of universities and public research centers to Communities where they did not previously exist; b) the effort made by public administrations. Some of them have implemented or are implementing concerted actions to promote research; and c) the application of European Union Structural Funds to R&D infrastructures in Spain since 1989.

– OCYT Science and Technology Office, falling under the Ministry of the Presidency in February 1998, which is a supporting institution for the CICYT. This office will be the support unit for planning, co-ordinating, monitoring and evaluating science and technology activities of the different ministry departments and public entities. It will also carry out all tasks necessary for co-ordination with the Autonomous Communities, as well as the co-ordination and monitoring of international R&D programmes.

– General Council of Science and Technology. This is the institution in charge of promoting and co-ordinating the activities among different Autonomous Communities and ensuring coherence among technological objectives in national and regional policies.

– The National Assessment and Prospective Agency (ANEP). This depends on the CICYT and is in charge of evaluating the scientific-technical quality of the financing applications presented to the PN I+D.

– The CIS (Sociological Research Centre) falls under the Ministry of the Presidency and its task is to study Spanish society, mainly by means of survey supported research. It is Spain's most important reference for public opinion studies.

– The CICYT has appointed the Centre for Industrial Technological Development (CDTI), depending on MINER, as its interlocutor with Spanish companies, in order to establish the most appropriate financial support mechanisms for the pre-competitive research projects, labelled Co-ordinated Projects.

– The Advisory Council for Science and Technology (CACYT) is in charge of promoting the participation of economic and social agents in the drafting, monitoring and assessment of the PN I+D. It is currently chaired by the Minister of Industry.

– The OTRI-OTT network (Results of Research Transfer Office and Technology Transfer Office) was established as an instrument to promote the presence of companies in scientific tasks, to articulate the science-technology-industry system and to make the scientific environment more dynamic. Each PRC has its corresponding OTRI.

– Among the public research centres, the most important one is the Higher Council of Scientific Research (CSIC), depending from the MEC, and the universities. It carries out most of the research, and also takes part in the selection of scientific and technological objectives in national and regional policies. It also collaborates in the counselling and management of different aspects of the PN I+D. The CSIC has a research centre network, each branch of which is devoted to different issues.

– The Ministry of Economy has a specific department, the Directorate General of Small and Middle-sized Enterprises Policy, which manages support to SMEs with a horizontal character. These decision-making powers belonged up to 1996 to the Institute of Industrial SMEs (IMPI), which has been eliminated. The activities of the IMPI have focused on providing companies with information, capital risk actions, promoting co-operation between Spanish and community companies and intervening in the SME guarantee system.

National R&D co-ordination mechanisms. The CICYT has promoted mechanisms to facilitate smooth exchanges among the different agents, to valorise R&D developed in public centres and to promote R&D in companies, preferably by means of their co-operation with public research.[10]

In order to improve the articulation of the science–technology–industry system, the 3rd National R&D Plan established a new '*National Programme for the Promotion of the Articulation of the Science-Technology Industry System*'. Its purpose is to promote the articulation of the scientific, technological and production environments and to promote an effective orientation and use of scientific and technological know-how and capacities by productive sectors and society as a whole.

On the other hand, the National Programmes of the National R&D Plan are already a co-ordination action in and of themselves, since they articulate the R&D groups around the priority objectives which include to a great extent sectoral initiatives and those of public and private entities.

Another co-ordination mechanism is the ATYCA initiative which serves as an umbrella that covers all actions related to the improvement of industrial technology, quality, security and environment implemented during the past years by the central administration.

6.3.3   Objectives and instruments

6.3.3.1   Science and technology policy
Since 1986, Spanish research policy has been based on the Law for 'Promotion and General Co-ordination of Scientific and Technical Research', known as the 'Science Law'. This law intended to correct the basic errors of the Spanish science and technology system, moving towards a socioeconomic, objective-based model that established the instruments required to define more precisely the priority work lines in R&D. The law used two main instruments for this purpose: establishment of a National

---

[10] The co-ordination mechanisms are as follows: the OTRI-OTT Network as interface structure of the science-technology-industry system; the Research Result Transfer Stimulation Programme (PETRI actions); joint projects agreed by companies and the OPI; and supports to the exchange of research staff between companies and OPI.

R&D Plan (PN I+D) since 1986, and definition of a common framework for the different public research centres.

The PN I+D pursues the correction of the historical deficiencies of the Spanish Science–Technology–Industry System by means of the promotion, co-ordination, and planning of R&D activities. The goal is to eliminate the barriers existing between basic and applied research, or between science and technology, and to strengthen the role played in research by universities and to co-ordinate the R&D centres depending on the ministry.

Currently the first and second stages of the PN I+D have been completed and the third stage, corresponding to the 1996–1999, period is underway. The last one sets out four main guidelines for core actions: creation of infrastructure; qualification of new researchers; development of new research projects; and fund allocation to companies and industrial development plans.

One of the Plan's objectives is to capitalise the science–technology–industry system and for that purpose the implementation of the Plan has meant the allocation of additional funds in the National Budget. Furthermore, the National Plan mobilises both public and private resources, thus producing a multiplying effect on the resources falling within its purview. This happens because most of the actions included in the Plan imply the investment of additional resources by the beneficiaries of the grants, either in the way of direct co-financing or as a share of the general operation costs or the labour costs of the staff implementing those actions.[11]

The CICYT is responsible for planning, co-ordination and assessment of the PN I+D. It is also in charge of co-ordinating research activities of the different ministry departments and government entities in implementing the National Plan, as well as of the co-ordination and monitoring of international R&D programmes with Spanish participation. It is intended that these objectives should be met by means of the following instruments:

- *Promotion and planning instruments*: R&D projects and special actions: Scientific–Technical Infrastructure: Integrated Projects; Strategic Mobilisation Projects; Training Actions for Research Staff; Setting Up and/or Reorientation of Research Teams.
- *Co-ordination Instruments*: Sectoral co-ordination and co-ordination with the Autonomous Communities.
- *Articulation Instruments*: National Programme for the Promotion of the Science–Technology–Industry System Articulation.
- Financial instruments: subsidies, awarded to public and private non-profit entities and reimbursable grants to the subsidised activity, which may

---

[11] It is estimated that for every *peseta* allocated by the Plan, there will be an additional 5 *pesetas* from other financing sources (3rd National R&D Plan).

under no circumstances exceed the limits established by EU regulations
for this type of grant.
– Tax incentives: In Spain the Corporate Tax Law is the one that defines
   which activities are eligible for tax benefits. It eliminates most stages of
   the innovation process, focusing almost exclusively on the research and
   development stages.

The Plan organises these activities in the way of scientific-technological
programmes that can be grouped in different types:
– *National programmes:* These are drafted by the CICYT and can integrate
   sectarian initiatives, both public and private ones.
– *Sectoral programmes:* These are drafted by the ministry departments
   involved and by other governmental entities.
– *Programmes of the Autonomous Communities:* These are financed totally
   or in part with national government funds and can be included in the
   National Plan.
– *National research staff training programmes.*

6.3.3.2   Industry research and technology policy

A new concept of industrial policy was conceived in Spain in the early
1990s. Traditionally the country had adopted a passive position, which was
subordinated to the imperatives of the macro-economic objectives
established by the Ministry of Economy and focused on the re-conversion of
industrial sectors in crisis. Then, the government became aware of the need
to set up new active industrial policy instruments based on the market. These
instruments tended essentially to promote the dissemination of technological
innovations and to modernise small and middle-sized enterprise in order to
generate new competitive advantages. It also assumes the need for a greater
co-operation among the main agents involved (Administrations, companies,
universities, etc.). Spain's integration into the EEC and the subsequent
assumption of the community's general industrial policy guidelines greatly
contributed to this change.
   The novelty is the search for the interrelation between industrial policy
and competition policy. Thus, the foundations of industrial policy in Spain,
issued by the Ministry of Industry and Energy (MINER) and its autonomous
bodies, are to strengthen competition policies and the definition of a new
paradigm of industrial policy based on the competitiveness, on intangible
assets and on innovation as the pillars of industrial development and job
creation within the sector. The strategy has moved from sectoral conceptions
based on subsidies or public contributions to the adoption of strategies that
promote innovation (MINER, 1997b).

More specifically, MINER's technological and innovation policy focuses on the achievement of the following objectives. (MINER, 1997b):
– To articulate more effectively the science–technology–industry system, promoting corporate R&D especially amongst SMEs and placing special emphasis on co-operation with public centres and the research associations and extending technology dissemination and transference.
– To disseminate and promote an innovation culture that promotes entrepreneurial attitudes in the society.
– To establish a new legal, regulatory, and financial framework that favours innovation. For this purpose different activities have been commenced, such as the adaptation and simplification of the R&D legal and industrial copyright frameworks; the administrative simplification of company constitution; the tax treatment of innovation activities and the increase of the R&D financing instruments.

Among the instruments established in the field of technological and industrial policy we should point out the ATYCA initiative, the SME Industrial Development initiative 1994–1997, the projects financed by the Technological and Industrial Development Centre (CDTI) and the National Programme of Advanced Production Technologies, which is included in the 3rd National R&D Plan 1996–1999.

Plan to Support Industrial Technology, Safety and Quality, 1997–1999 (ATYCA)
ATYCA was launched in January 1997 and it succeeded the Industrial Technological Action Plan (1st PATI 1991–1993 and 2nd PATI 1994–1996)[12] and the National Industrial Quality Plan 1994–1997 (PNCI). The ATYCA initiative intends to promote innovation in the corporate environment, with a special emphasis on the SMEs, in order to increase their competitiveness by means of the development and introduction of horizontal and disseminating technologies, and the incorporation of quality and security into all industrial processes. It was drafted with the purpose of co-ordinating and grouping all the support and subsidies provided by the Ministry of Industry and Energy in terms of industrial technology, safety and quality and environmental industry

---

[12] The PATI was established to respond to the inadequate effort in technological development and to the need to incorporate advanced technologies in Spanish industrial companies that depended technologically on foreign countries. The specific objectives of the PATI were to increase the number of accredited companies and certified products, to encourage consumer demand for safer and more reliable products and to reinforce the supervision and control measures for industrial products according to the patterns established by the European Union.

in order to establish a reference point for all companies searching for support for their technological and industrial quality projects.

Thus, the ATYCA initiative includes the objectives and instruments of the Environmental Industrial and Technological Programme 1990–94 (PITMA), the purpose of which is to promote R&D and the technological adaptation of companies to the environmental regulations in force and to promote the co-ordination of a national environmental industry supply. Other programmes that have been incorporated into ATYCA are the Technological Capacity Promotion Programme (aimed at the construction of infrastructure and the extension of innovation in the less favoured objective 1 regions) and the actions of the Technological and Industrial Development Centre (CDTI).

ATYCA foresees the allocation between 1997 and 1999 of subsidies amounting to €398.49 million to corporate projects. They will be divided into two groups of projects with separate objectives and budget allocations:
– The Industrial Technology Promotion Plan (PFTI) focuses on the promotion of specific technologies, with horizontal activities in order to improve infrastructures, training, and support systems for corporate innovation.
– The Quality and Industrial Safety Programme (PCSI) focuses on the introduction of quality management systems in companies and at the same time on the promotion of certification and eco-labelling in order to increase exports of Spanish industrial products. These were the objectives pursued by the former National Industrial Quality Plan (PNCI) 1994–1997.

ATYCA has also incorporated a technological area devoted to the development of industrial design. It is also complementary to the international R&D actions in which Spain participates, such as Eureka, Iberoeka, the European Space Agency, and the EU Framework Programme.

SME Industrial Development Initiative 1994–1999
This initiative is managed by the Ministry of Economy and Treasury in collaboration with the Autonomous Communities. Its purpose has been to reinforce the competitiveness of Spanish industrial SMEs within growth and employment policy. At the same time it has simplified the complex institutional web of supports and extended the information and use by SMEs of public support systems.

It was approved in 1994 and introduced in 1996. It defines the measures and actions aimed at achieving an improvement of the competitiveness of industrial SMEs, as well as the guidelines and working patterns for the Public Administrations.

Projects financed by the Technological and Industrial Development Centre (CDTI)

The CDTI is an entity that falls under the Ministry of Industry and Energy, founded on August 5th 1977. Its functions are defined by the Science Law and its objective is to help Spanish companies raise their technological and R&D standards. For this purpose it uses preferential financial instruments (zero interest loans) that support the implementation of research and development projects. Furthermore, the CDTI promotes the participation of Spanish companies in international R&D programmes and supports those that opt for internationalising the technological side of their business.

The CDTI assesses and finances following type of projects:
- Concerted and Co-operative Projects: Pre-competitive research projects, the results of which are not directly marketable. They incorporate a high technical risk and are implemented by companies in collaboration with Universities or Public Research Centres if they are concerted projects, or with Technological Centres if they are co-operative ones, in the areas envisioned under the National R&D Plan. The CDTI grants zero interest loans from the budget of the National R&D Plan Fund (included in the budget of the Ministry of Education and Culture of the Central Administration).
- Technological Development Projects: These are applied R&D projects implemented by companies. A Technological Development Project might entail the creation or improvement of a product or a productive process and must have a short or mid-term effect on the company. The CDTI finances between 40% and 50% of the total cost by means of low-interest loans.
- Technological Innovation Projects: These are oriented towards the incorporation and active adaptation of new technologies in the company rather than their development. They not only entail the replacement of isolated equipment or technological elements but also imply greater changes in the production system and in corporate organisation. The CDTI finances up to 25% of the total cost of these projects by means of soft loans.[13]

---

[13] Technological development and technological innovation projects are financed in objective 1 regions by means of reimbursable grants without interest with the participation of the ERDF through the Global Subsidy ERDF-CDTI.

National Programme of Advanced Production Technologies
This programme, incorporated into the 3rd National R&D Plan aims at the improvement of production processes by means of the incorporation into them of advanced information processing, automation, and organisation techniques. The 4th European Union Framework Programme follows this same orientation, especially in the ESPRIT programme.
The general objectives of the programme are:
– To promote technological evolution in the field of production technologies taking onto account the new trends reflected in the European Union Framework Programme
– To promote research in new fields related to production technologies.

The specific objectives of the programme are based on a global perspective of the production process, which considers both the product life cycle and the support technologies to the production process.

6.4   Policy interactions

Research and technological innovation are essential elements to create competitive advantages and sustainable development. Furthermore, technological development is critical for the long term conservation of resources and the improvement of the environment and that is why it is so important to favour innovation processes that effectively meet specific demands (technological, economic, and social, as well as environmental ones) in the regions. Moreover, the complexity and specificity of new technologies applied to environmental management suggest the need to set up a government strategy specifically oriented at dynamic innovation processes with environmental purposes.

Thus, the consideration of the environment as an innovation factor has caused the starting up of certain programmes by the central administration both in the field of research (National R&D Programme on Environment and Natural Resources) and in the field of industrial technology (ATYCA initiative and SME Industrial Development initiative), both of them oriented to the environment, so they include combinations of environmental and technology policy instruments. Three ministries are responsible for the implementation of these policies: The Ministry of Education, the Ministry of Industry and Energy and the Ministry of Economy.

In conclusion, there is a clear relationship between technological innovation and environmental preservation. The increasing environmental awareness of companies, administration and society in general, leads to the development and incorporation of new and more environmental friendly technologies in industry (BATs). For this reason, co-ordination between technological policy and environmental policy is necessary to achieve the principles of sustainable development. This co-ordination is still inadequate in Spain and more efforts are required to increase the integration between both policy systems.

National R&D Programme on Environment and Natural Resources
This is one of the Programmes envisioned in the 3rd National R&D Plan. It started in 1992 and resulted from the merger of several programmes of the 1st National Plan: the ones for the Preservation of the National legacy and for Environmental Degradation Processes, for Forestry Resources and for Sea and Geology Resources.

The general objective of the Programme is to study the pressure that human activities place on the environment and the possibility that the changes generated by this human pressure may be irreversible, at least in the short and medium term. It considers that research must be structured taking into account the effects of human activities on the environment in all its aspects: the ones related to systemic changes (global change and environment), cumulative ones (air, water, and soil pollution), correction methods (technologies) and the effects of human activities on society itself (economic and social impacts). Thus, it covers four priority areas: global change and environment; physical-chemical processes and environmental quality; technologies for the environment; and environment and social and economic impacts.

The specific objectives of the programme are:
– To reduce the differential between the Spanish environmental industry and the European one in the field of environment and sustainable development, climate change, biodiversity, and desertification, with emphasis· on economic growth and employment.
– To provide the R&D system with the necessary continuity, supporting problem resolution or projects presented in earlier stages of the programme.
– To encourage and improve the participation of public research centres and companies in international R&D programmes.

The Programme is articulated around sectoral programmes that cover the main priority areas in environmental issues in Spain such as: water quality and optimal use, waste management, the fight against desertification and loss of fertile soil, impacts of climate change, preservation of biodiversity and the quality of the urban environment.

However, some environmental aspects and problems are covered by other programmes of the National R&D Plan such as those of Sea Science and

Technology, Biotechnology, Food Technology, Rural Research and Development and Water Resources in Climate R&D.

There are three differentiated areas in the Environmental Technology chapter:
1. Technologies for environmental surveillance, both 'on site' and by tele-detection. Its main purpose is to draft and improve measures for the analysis and control of pollutants' emissions in order to watch over and predict environmental changes.
2. Technologies to reduce pollution, both in conventional processes and in polluted area. There is a distinction between the projects that present new concepts, new technologies or their environmental application and those that develop, improve or optimise conventional technologies.
3. Cleaner technologies. In this field priority is given to conception and assessment projects at laboratory or pilot plant scale that enable the establishment of the feasibility of technological changes, process improvements and environmental optimisation actions. Likewise, priority is given to projects aimed at testing alternatives. Projects related to technologies for waste recycling and re-use are also covered.

Technologies for the environment of ATYCA Initiative (1997–1999)
One of the areas where ATYCA operates by means of the Industrial Technology Promotion Programme is the area of technologies for industrial environment, and more specifically by the implementation of the following:
– Industrial prevention practices.
– Rationalisation of raw materials and valorisation of waste.
– Optimisation of traditional production technologies.
– Cleaner technologies and use of improved technologies at a reasonable cost, especially in the industrial sectors affected by the EU Directive (96/61/CE) IPPC.
– Sustainable management systems and audits.

The objective of ATYCA, as far as eco-management and eco-auditing are concerned, is to promote a continuous improvement of corporate behaviour towards the environment by means of the implementation of environmental policies, programmes and management systems (certification of the ISO 14001 norm, for example); the systematic and periodical review of their commitments with the environment (audits) and the information to the public of their situation regarding the environment.

Furthermore, the ATYCA initiative supports companies in their bid to obtain the European ecological label by granting support for actions, such as product lifecycle and industrial production technology introduction research studies, that allow more environment friendly products to be manufactured.

However, the implementation of these kinds of instruments is not as well developed in Spain as in other countries. For example, regarding to the IPPC directive, we are still in the phase of establishing task forces to inform industries about the scope for technological improvement (we cannot forget that the period for implementing this directive finishes at the end of 1999).

6.5   Conclusions

6.5.1   The new role of public administrations

In general terms, the SCTI shows a great separation between the scientific and the production environments. The production environment lacks the facilities and strategies to tackle the necessary innovation in products and processes. On the other hand, the technological or technological-industrial environment is not sufficiently developed to be able to benefit and to translate effectively the national or foreign scientific developments into new industrial processes and equipment for their use in the production environment. The scientific environment has adopted too conservative a position, waiting for proposals from the production environment.

For all the above reasons, public administrations are becoming aware of the need to reinforce and improve the interface between the Spanish research system and the business world by promoting the creation of consortia between companies and research centres. However, the mechanism is quite complex and this prevents the industrial sector from participating on a regular basis, something that should be an essential part of a strategic approach to competition in markets that are becoming bigger every day (COTEC Report, 1997).

On the other hand, the Spanish decision-making process is quite complex. This is shown by the number of bodies involved in the process and the changes experienced by the different administrations involved, resulting from the organisational changes aimed at improving efficiency and effectiveness. This situation has had clear implications in the operational outcome of innovation, industrial and environmental programmes launched by the Administration. Thus, for example, the Environmental Industrial and Technological Programme (PITMA) was managed by the Ministry of Industry and Energy (MINER) until 1996, when the Ministry of Environment was created. This programme was substituted by the ATYCA Initiative (which inherits its principles and objectives in the section on environmental technologies) which is again managed by MINER.

Another example can be found in the Industrial Development SME Initiative. This initiative was managed by the Industrial Small and Medium-sized Industrial Enterprise Institute (IMPI) until it disappeared in 1996. Since then the Directorate General of SME Policy of the Ministry of Economy and Treasury has been responsible for its management.

All this restructuring has caused the generation of a certain administrative 'stress' in the normal implementation of the actions. Additionally this makes the corporate decision-making process more difficult, due to their mistrust of the Administration, the excessive number of bureaucratic procedures and the absence of clear information on who is responsible for the programmes.

6.5.2   Increase of industrial competitiveness and reinforcement of the S&T system

The fact that the Spanish economy is immersed in the new globalization dynamic means that industry has to face increasing competitiveness. In this competitive scenario Spanish companies must increase technology and innovation investments and their preferential objective should be the reduction of their environmental impact in considering potential business strategies.

The quantitative measures used in analysing the competitiveness of the industrial sector of an economy are the variables referring to productivity and unit labour costs. If we analyse these variables, labour productivity has followed a constantly growing evolution; however, there are sectors that have higher competitiveness levels than others (such as hydrocarbons, refineries, gas, electric energy, and petrochemical). Regarding labour costs in Spanish industry, they are below the EU average; however, they are higher than the labour costs in the less developed countries that are its main competitors, especially for little-differentiated final consumption goods.

Despite the efforts made by companies to improve their competitiveness, it is necessary to enhance them yet further to bring Spanish industry to a competitive position in the global market. These efforts must be focused on the increase of industrial R&D and the adoption of business structures that respond to the new market requirements, moving from local approaches to global ones. Thus, Spanish industry must allocate financial resources to the incorporation of clean technologies into their production system they must adopt more environmental friendly practices and behaviour, introduce environmental management systems and in general, incorporate in industry these new requirements, which must be promoted both by private business initiatives and by the Public Administration.

The main conclusion is that the innovation system, on the one hand, is in need for effective co-ordination among institutions and agents - especially the central administration, the Autonomous Communities, the innovation support infrastructures, and the business co-operation networks that intervene in the science-technology-industry system – in order to secure an adequate distribution of the resources allocated to innovation.

On the other hand, the effort of companies aimed at improving their competitiveness must be promoted. One of the main ways is the necessary simplification of bureaucratic procedures and administrative formalities.

6.5.3   Environmental pressures and consumer concerns

As we mentioned above, environmental conservation is nowadays a factor that influences competitiveness in economic activities. Spanish companies used to consider the environmental factor just like any other expense on their balance sheet, but currently it is considered as an element that generates economic and social profitability.

Due to the new consumer trends towards the demand of environmentally acceptable products (the so-called 'green consumer'), companies must adapt their production planning towards the development of production techniques that incorporate the environmental variable. There are different social agents promoting these actions. We should highlight the participation of labour unions that in their double position of workers and consumers condition the application of environmental policies within the companies.[14]

Regarding clients, consumers, and public opinion in general, we must emphasise that the general population plays an essential role in environmental conservation and in the business implication this awareness-raising process entails. Most Spanish citizens think that environmental conservation is an urgent, priority issue, while only 12% opt for economic development rather than environmental conservation (MINER, 1997a). The survey conducted by MINER reaches the conclusion that most citizens are willing to assume some of the costs in exchange for improving the environment, which indicates a high degree of social involvement.

All of this puts increasing pressure on Spain's productive industries. Companies have not just to adapt to environmental regulations (which are becoming increasingly stringent) but must also enter into voluntary agreements to assume better behaviour towards the environment in order to adapt to consumers' environmental requirements.

These demands make technological developments that generate new products and processes become associated with innovation agents. Thus, over the last few years we can detect an increase in the number of international collaborations, with a greater role played by companies,

---

[14] In Spain the majority labour unions, U.G.T. and CC.OO. are organised in environmental areas with clearly defined objectives.

especially in the so-called dissemination technologies. Terms such as 'techno-globalisation' have been recently coined to reflect this new trend.

6.5.4   Company strategies and networking

Spanish industrial companies have mostly adopted in the field of innovation a business strategy, i.e. they aim at the technological leadership within their own activity sector.[15] We should be more precise about this statement since there is a lack of human, financial, and material resources devoted to innovation, aside from the low degree of specialisation of the technological generation and transfer processes used, means that strategy adoption can be characterised by imitation rather than by innovation leadership. In this field both public administrations and private companies should enter into agreements aimed at the promotion of greater investments in innovation in those fields defined as priority ones (environment, quality, technologies, etc.).

Regarding environmental issues, the business strategy adopted by most Spanish industrial companies is one of 'compliance'.[16] This position entails that compliance with environmental regulations does not provide a competitive advantage compared to other companies since most of them will have a similar environmental behaviour. However, recent studies (IDR) indicate a change towards the incorporation of the environmental variable as it means a competitive advantage. In conclusion, industry must direct its business strategy to the achievement of, at least 'compliance plus', or even to 'environmental excellence'.

6.5.5   Job implications

In general terms industrial employment grew in 1996 (0.6% increase as compared to the previous year). The growth of the employment variable in manufacturing industry experienced in 1996 an increase in the group with a high technological level (after four years of continuous reduction), an increase in sectors with a middle technological level and a decrease in those activities with a low technological component. Employment trends show a reorientation of production activities towards sectors with a higher technological component, which forces industrial employment to evolve towards higher qualification and production specialisation in activities with

---

[15] *La industria española ante el proceso de innovación* (Spanish industry in the face of the innovation process) MINER (1997).

[16] Using the classification made by Roome (1992) that identifies the following types of environmental strategies regarding the environment: 'non-compliance', 'compliance', 'compliance plus', 'environmental excellence' and 'leadership'.

higher technological contents.[17] However, the persistence of high structural unemployment rates has caused the government to launch policies giving incentives to job creation.

In this regard, the application of environmental policies has created a controversy about how good they are, since they might be in contradiction with the policies aimed at economic growth. The analysis done by the Spanish industry on the effects the incorporation of the environmental variable has had on business planning strategies shows positive results: this environmental trend implies new training needs within companies, aside from being a new focus for the creation of jobs with high qualification levels.

In conclusion, we can state that the implications of technological and promotion policies in the environment field also cause in the Spanish case the creation of new jobs. Therefore, environmental conservation and technological innovation are not in contradiction with economic growth.

REFERENCES

Aragón Correa, J.A. (1998) *Empresa y medio ambiente. Gestión estratégica de las oportunidades medioambientales*. Granada: Comares.

Comisión de las Comunidades Europeas (1992) *V Programa Marco en Medio Ambiente "Hacia un desarrollo sostenible"*. Unión Europea.

Comisión de las Comunidades Europeas (1998) V Programa Marco en I+D.

Comisión Interministerial de Ciencia y Tecnología (1999) *III Plan Nacional de I+D. 1996-99*.

Conesa Fernández-Vitora, V. (1995) *Auditorias Medioambientales. Guía metodológica*. Madrid: Ediciones Mundi- Prensa.

Conesa Fernández-Vitora, V. (1996) *Instrumentos de la gestión medioambiental en la empresa*. Madrid: Ediciones Mundi-Prensa.

Fundación COTEC para la Innovación Tecnológica (1997) *Documento para el Debate sobre el Sistema Español de Innovación*. Madrid: Fundación COTEC.

Fundación COTEC para la Innovación Tecnológica (1998a) *El sistema español de innovación. Diagnósticos y recomendaciones. Libro Blanco*. Madrid: Fundación COTEC.

Fundación COTEC para la Innovación Tecnológica (1998b) *Segundo análisis del tratamiento de la innovación tecnológica en la prensa española (1996-97)* Madrid: Fundación COTEC.

Fundación COTEC para la Innovación Tecnológica. Informe COTEC. Varios años. Madrid: Fundación COTEC.

Fundación Entorno, Empresa y Medio Ambiente (1998) *Libro Blanco de la gestión medioambiental en la industria española*. Madrid: Ediciones Mundi-Prensa.

García Delgado, J.L (Director), Myro, R. Y., Martínez Serrano, J.A. (eds.) (1997) *Lecciones de Economía Española*. Madrid: Editorial Civitas.

---

[17] Report on Spanish industry 1996–1997 MINER (1997).

Instituto de Desarrollo Regional. (1998a) *Estudio de casos de políticas de Fomento para el desarrollo de la industria de bienes y servicios medioambientales"*. Sevilla : IDR (mimeo).

Instituto de Desarrollo Regional. (1998b) *Programas europeos para el desarrollo de la industria medioambiental.* Sevilla : IDR (mimeo).

Instituto de Desarrollo Regional (1999a) *"La oferta regional del sector medioambiental y la demanda industrial de bienes y servicios medioambientales en Andalucía"*. Sevilla : IDR (mimeo).

Instituto de Desarrollo Regional (1999b) *"La orientación de las actividades de innovación de los grupos de investigación andaluces en materia de medio ambiente"*. Sevilla : IDR (mimeo).

Instituto de Desarrollo Regional (1999c) *"Medio ambiente y desarrollo regional. La contribución de la industria de bienes y servicios medioambiental al desarrollo regional. Una especial referencia al caso de Andalucía"*. Sevilla : IDR (mimeo).

MINER (1997a) *La industria española ante el procesos de innovación.* Madrid: Secretaría General Técnica. MINER.

MINER (1997b) *Informe sobre la Industria Española, 1996-97.* Madrid: Secretaría General Técnica. MINER.

Seoanez Calvo, M. (1995) *Ecología Industrial: ingeniería medioambiental aplicada a la industria y a la empresa.* Madrid: Ediciones Mundi-Prensa.

LIST OF ABBREVIATIONS

AENOR. Spanish Normalisation Association

ANEP. National Assessment and Prospective Agency.

ATYCA. Plan to Support industrial Technology, Safety and Quality

BATs. Best Available Technologies.

CACYT. Advisory Council for Science and Technology.

CAMA. Environment Advisory Council

CDTI: Technological and Industrial Development Centre.

CICYT. Inter-ministry Commission for Science and Technology.

CIS. Sociological Research Centre.

CSIC. Higher Council of Scientific Research.

EEC. European Economic Community.

ERDF. European Regional Development Fund.

EU. European Union.

FEDIN. Spanish Federation of Research Associations.

FEDIT. Spanish Federation of Innovation Entities.

IDR. Institute for Regional Development.

IMPI. Industrial Small and Middle-sized Enterprise Institute.

LRU. University Reform Law (1983).

MEC. Education and Science Ministry,

MINER. Ministry of Industry and Energy.

MOPTMA. Ministry of Public Works, Transport and Environment.
NGOs. Non Governmental Organisations
OTRI-OTT. Research Transfer Results Office.
PATI. Industrial Technological Action Plan.
PCSI. Quality and Industrial Safety Programme.
PETRI. Research Result Transfer Stimulation Programme.
PFTI. Industrial Technology Promotion Plan.
PITMA. Environmental Industrial and Technological Programme.
PN I+D. National R&D Plan.
PNCI. National Industrial Quality Plan.
R&D. Research and Development.
RELE. Spanish Network of Testing Laboratories.
SCTI. Science-Technology-Industry System.
SME. Small and Medium Enterprise.

Chapter 7

Environmental Policy and Environment-oriented Technology Policy in the United Kingdom

JOHN F. GRANT AND NIGEL D. MORTIMER
Resources Research Unit, School of Environment and Development, Sheffield Hallam University, United Kingdom

7.1   Introduction

Currently, environmental activities in the United Kingdom are usually set in the context of sustainable development. Many diverse actions are required to achieve the practical realisation and successful implementation of sustainable development. Amongst the numerous challenges which this presents, fundamental progress is needed to develop and disseminate technological innovations which will reduce the impact of human activities on the natural environment. An appropriate framework for fostering such innovation is, therefore, an essential component of action for sustainability. In order to consider suitable frameworks for innovation in the United Kingdom, it is necessary to examine current UK environmental policy and environment-oriented technology policy and determine whether they are effective in creating relevant conditions to assist technological innovation. Using this basis of policy analysis and actual experience from companies and organisations involved in the innovation process, it is then possible to establish and recommend approaches which can encourage the development and commercial application of environmentally-beneficial technologies for future sustainability.

This chapter presents the results of the analysis of the operation of environmental policy and environment-oriented technology policy in the United Kingdom. section 7.2 describes the historical context of UK

Geerten J.I. Schrama and Sabine Sedlacek (eds.) Environmental and Technology Policy in Europe.
Technological innovation and policy integration, 197-224. © 2003 Kluwer Academic Publishers. Printed
in the Netherlands.

environmental policy, the main participants in the implementation of this policy and the key policy instruments currently in use. Similarly, section 7.3 considers the historical context of technology policy in the UK and examines the importance of the Foresight Programme as a major policy instrument for assisting technological innovation, in general, and advances in environmental technology, in particular. The broad framework for promoting innovation in the United Kingdom is discussed in section 7.4, especially in relation to UK policy on competitiveness and its subsequent implementation. Finally, in section 7.5, conclusions are drawn as regards the characteristics of UK environmental policy and environment-oriented technology policy.

7.2   Environmental Policy

7.2.1   Historical context

In the United Kingdom, present awareness concerning the environment began during urbanisation in the 19th century with the Public Health Acts (1848 and 1878) and the Alkali Act of 1864 (which limited industrial emissions). More recently, a Royal Commission on Environmental Pollution was established with a duty to provide advice and information to the UK government on environmental issues which were becoming increasingly apparent to the scientific community, the public and politicians during the 1970's. Over this decade, the Commission reported on numerous environmental problems, but in relation to specific industries or types of pollution rather than the environment as a whole.

In 1987, 'Our Common Future', commonly called the Brundtland Report (World Commission on Environment and Development, 1987), changed the general point of view, especially the perspective of politicians in the United Kingdom. This report set an agenda for discussion about the relationship between economic growth and protection of the environment, encompassed by the concept of sustainable development. This commitment was followed by further developments in the European Union, generally, and in the United Kingdom, specifically. In the European Union, the 'Fifth Action Plan: Towards Sustainability' (European Commission, 1993a) identified shared responsibility and environmental information as the key elements in enabling citizens to participate in environmental protection and sustainable development, and as a means of exercising consumer preference for given products and producers.

Prior to publication of the Brundtland Report, mainstream politicians in the United Kingdom had a reputation for regarding general environmental concerns as minor issues with respect to national public voting intentions.

Local environmental issues were seen as important to local politicians, of course, but wider environmental involvement was chiefly viewed as the preserve of fringe political movements, such as the Green Party. The shock of the 1989 European Parliament elections, which resulted in a sudden increase in votes for the Green Party, albeit from a very low base, caused a fundamental change in attitudes amongst the politicians of the Conservative government and the opposition Labour Party. This resulted in the publication of 'This Common Inheritance' in 1990 by the Conservative government, as the first comprehensive statement of environmental policy in the United Kingdom (Department of the Environment, 1990). It contained references to all the main areas of concern, such as greenhouse gases, land-use conservation, pollution control and its effects on air, water and land. The cornerstone principle of this document was 'stewardship', which means everyone has a responsibility to act as stewards of the environment and preserve it for future generations. Both UK and EU policy documents emphasise the importance of public participation in the protection of the environment, with the recognition that environmental protection cannot be secured by government action alone.

7.2.2   Environmental policy instruments

In the United Kingdom, the most important legislation for the direct implementation of environmental policy is the Environmental Protection Act 1990. The main thrust of the Environmental Protection Act (EPA) is prevention rather than cure; by controlling emissions to all media of the environment. This approach is characterised by Integrated Pollution Control (IPC) and subsequent implementation in the United Kingdom of the EU Directive on Integrated Pollution Prevention and Control (IPPC) in recent times (European Commission, 1996). The EPA makes it an offence to pollute with no need to prove intention or negligence, although harsher sentencing will be made if such contributory attitudes or behaviour are proven. It should also be noted that, in the United Kingdom, it is the 'state of the environment' which is the prime concern rather than the 'actual' emissions from a specific plant or process. This allows a more flexible approach, taking into account the site, situation and adjacent land-use issues in a specific area.

There are a number of important organisations involved in the implementation of environmental policy in the United Kingdom. These include the Environment Agency (EA), local authorities, and private industry. The most prominent organisations are the Environment Agency, which operates in England and Wales, and its equivalents in Scotland, the Scottish Environment Agency (SEA), and the Environment and Heritage

Services and Environmental Protection in Northern Ireland. These Agencies were set up to regulate and police environmental policies in the United Kingdom. The Act of Parliament establishing the Environment Agency was passed in November 1994 and, following publication of the Environment Act, the Environment Agency became operational on 1 April 1996. The Environment Agency is an amalgamation of three previously separate organisations:
– Her Majesty's Inspectorate of Pollution (HMIP), addressing air pollution,
– National Rivers Authority (NRA), concerned with water pollution, and
– Waste Regulation Departments at County Council level, related to waste disposal, chiefly in landfill sites.

The theoretical approach of the Environment Agency is to achieve a holistic and integrated system which can establish cross-medium pollution and seek to:
– prevent,
– remediate,
– compensate, and
– convict.

IPPC is now an integral component of the approach adopted by the Environment Agency. This is a mandatory instrument of UK environmental policy. In its application, it focuses on the 'best practicable environmental option' (BPEO) for addressing the prevention and control of pollution from any given process or activity. This involves determining the most appropriate media which should form the destination for a specific pollutant. Having established this, the subsequent response is based on the application of the concept of the 'best available techniques not entailing excessive cost' (BATNEEC). Both BPEO and BATNEEC are regarded as the core qualitative principles of the use of IPC in the United Kingdom when originally introduced in Part 1 of the Environmental Protection Act 1990. This approach has continued with adoption of the EU Directive on IPPC. As such, the approach has been criticised for an absence of clear and comprehensive statutory standards of environmental quality (Gouldson and Murphy, 1998).
    The guiding principle of the Environment Agency is sustainable development. The basic difference apparent in approaches to pollution control compared to the rest of the European Union is that the United Kingdom tends to consider the environment as a whole, and looks at its measured 'quality' rather than the application of individual emissions standards. Concern has been expressed regarding the effectiveness of this approach to environmental policy in the United Kingdom (Gouldson and

Murphy, 1998). In particular, it has been suggested there are three potential failings. First, there is apparent emphasis on the role of scientific certainty in environmental policy development which has resulted in possible dilution of the importance of the 'precautionary principle' in the United Kingdom. Second, it is argued that the style of implementing UK environmental policy relies too heavily on informally negotiated settlements. Third, it is stated that UK environmental policy 'has been influenced by free market doctrine'. Although these factors may have been influential during the earlier development of current UK environmental policy, there have been significant changes of direction and emphasis subsequently.

Previous negative conclusions concerning UK environmental policy have chiefly been based on an analysis of the apparent attitudes of the Conservative government between 1979 and 1997, and the actions of the regulatory organisations, mainly prior to their amalgamation into the Environment Agency. The interpretation of government attitudes towards the environment are principally based on two documents; 'This Common Inheritance' (Department of the Environment, 1990) and 'Sustainable Development: The UK Strategy' (Department of the Environment, 1994). However, the statements contained in these documents have been superseded by those of the current Labour government. A number of new policy statements have been issued since 1997 covering major environmental issues such as climate change, transport policy, and waste minimisation. There is now increasing acceptance of the need to set targets for environmental improvement. For example, the United Kingdom has now established clear and ambitious targets for reducing specific air pollutants, such as particulates and benzene (Department of the Environment, Transport and the Regions, 2000). Additionally, the operation of the Environment Agency has become increasingly effective as experience has been gained since 1996. Although the approach retains its basis of negotiation, there has been a significant shift in government policy towards strengthening the role of regulation of free market operations by official organisations such as the Environment Agency. This appears to be regarded as an essential development for a government which sees its relationship with private industry as one of partnership in the field of environmental policy.

As a further indication of better understanding of the required approach to environmental policy, integration became, at one time, a more important consideration in decision-making. This was characterised by the merging of three government departments, including the Department of the Environment (DoE), into the Department of the Environment, Transport and the Regions (DETR) in 1997. Unfortunately, a decision was made in 2001 to separate the DETR into the Department of the Environment, Food and Rural Affairs (DEFRA) and the Department for Transport, Local Government and the

Regions (DTLR). This divided environmental policy functions once again with, for example, the DEFRA being responsible for the Environment Agency and the DLTR handling planning issues (see Section 7.2.3). Moreover, UK environmental policy has been criticised for a lack of fundamental advances, especially by non-governmental organisations such as Friends of the Earth and Greenpeace. Such criticism has often concentrated on conflicts between different government policies, particularly on the economy, transport and the environment. However, it is clear that proper understanding of UK environmental policy and its subsequent influence can only be gained by appreciating its broader context, including the basic role of the established planning system.

7.2.3   Planning system in England and Wales

The UK planning system provides an established framework within which developments related to certain aspects of environmental policy and environment-oriented technology may occur. It provides a means of development control rather than a mechanism for promoting development directly, as may be the case in some other EU member states. The UK planning system rests on national legislation, currently the Town and Country Planning Act 1990, which sets the framework for activity over the whole of England and Wales. This is only a guide to the system. Whilst details of policy may vary between local authorities in England and Wales, the underlying system, expressed in terms of national policy guidelines, types of plan produced, consent procedures and legal details, are common to England and Wales in their entirety. Scotland and Northern Ireland operate under a related but slightly different system. Hence it is not correct to take the following description as valid for the whole of the United Kingdom. Land use planning is a function of government which is exercised in a hierarchical fashion. The central government department which carries ultimate responsibility for both the operation of local government and for land use planning is the DTLR.

Beneath central government there exists a sometimes complex arrangement of local authorities and other organisations, all of which have planning functions.[1] In most cases, a two tier structure exists in which counties provide the upper tier and districts the lower tier. The number of districts within a county varies, with the average being around six. Councils

---

[1] It should be noted that, since 1992, a Royal Commission on Local Government has been considering possible future changes to the definition and boundaries of local authorities in England, Wales and Scotland. Once the changes are finalised they will inevitably affect the operation of planning in some areas, particularly in terms of the types of plans produced.

of elected representatives exist at both levels and a broad range of functions are split between the different levels. Planning is exercised at both levels, although it differs in its nature and purpose as outlined below. All of the counties are known collectively as the 'shire counties'. Seven areas exist, all in England, where a different system operates. The areas are the former administrative counties of Greater London, Greater Manchester, Merseyside, South Yorkshire, Tyne and Wear, West Midlands, and West Yorkshire. In all of these areas, the county level of the hierarchy was abolished in 1986 and the majority of functions, including all planning functions, was vested in the districts. Thus, those district authorities which exercise a wider range of powers than the shire districts, are known as 'unitary planning authorities'.

A number of different area designations, set out for widely differing purposes, further complicate the picture. Of the ten national parks in England and Wales, two, namely the Lake District and the Peak District, have planning powers as unitary authorities. A wide range of other area designations can have implications for planning decisions, as discussed elsewhere (Grant, 1994). None of these designated areas, however, are controlled by bodies other than the county or district planning authorities within which they lie. Such definitions simply imply a greater or lesser amount of restriction on development depending on the reason behind the designation (for example, nature conservation or economic growth).

7.2.3.1 Main legislative instruments

Central government sets out policy on a wide range of land use related issues. The object of this policy is to assist local authorities in drawing up plans, provide guidance for decisions on development applications and ensure a consistent approach throughout England and Wales. There are three main vehicles employed. Firstly, the planning acts themselves are permissive in that they allow the relevant minister from time to time to introduce or modify statutory instruments. Secondly, central government produces circulars which seek to explain and advise on the application of acts and statutory instruments. Thirdly, the use of Planning Policy Guidance Notes (PPG) sets out central government policy on discrete issues. All of these different documents influence local authority plan-making and development control decisions. There is, however, no national plan.

At a local authority level a number of different types of plan are produced by these authorities depending on:
– the place in the local government hierarchy, and
– the purpose of the plan.

All shire counties have a statutory duty to produce a structure plan. This is a strategic document which looks to the long term (approximately fifteen years ahead). It is not site specific and may not contain Ordnance Survey based maps. It examines social and economic trends within the county and provides policies which are generally applicable over the whole county. Content usually ranges over subjects like housing, industry, employment, transport, nature and landscape conservation, agriculture, recreation, and tourism. The shire districts also have a statutory duty to produce local plans which cover the whole of their administrative area. These are detailed tactical documents which seek to interpret structure plan policies into site specific information, and local plan policies. As well as the district local authorities, the two national parks with full planning powers are under a statutory obligation to produce local plans. The content of all of the local plans so far discussed, in terms of subject matter, broadly reflects that of structure plans. Most will, therefore, specifically refer to issues of housing, industry, transport, conservation, or retailing. A separate class of local plans also exists which deals with specific topics or subjects. These are produced by counties and refer to mineral extraction and waste disposal, both of which are issues dealt with exclusively at county level.

The unitary authorities produce different plans from the shire counties and districts. Each district is charged with the statutory duty of producing a unitary development plan (UDP) which serves the dual function of structure plan and local plan for the whole of the district. It is divided into two parts which broadly equate to those types of plan. UDPs also include policies on minerals and waste disposal. There is no separate system of local plans specific to these topics in unitary authorities.

Although local authorities have limited powers to intervene directly in the development or implementation of environment-oriented technology policy (see Section 7.3), they can exert influence through the planning system. For example, as a result of the specification of land designations and approval of new development plans, which are guided by the need to improve urban air quality, private car usage can be discouraged and public transport can be promoted. Such action may create favourable conditions for the development and realisation of new transport technologies which can achieve even higher standards of environmental performance. However, the impact of such indirect involvement in environment-oriented technology policy depends crucially on the pro-active approach of individual local authorities.

7.2.3.2   Development control issues
The power to grant planning consent for development is vested in the state. A statutory definition of development which is all-embracing exists, namely:
"The carrying out of building, engineering, mining, or other operations, in, on, over or under land or the making of any material change in the use of

any buildings or other land" (Town and Country Planning Act 1990, Section 55)

Thus, any sort of physical development, including demolition or the change of use of buildings or land is technically 'development'. Whether this development requires the express grant of planning permission is normally decided by reference to the relevant statutory instrument. The General Permitted Development Order 1995 (GPDO) effectively exempts a large amount of minor development such as small changes to dwelling houses from the need for express planning permission. Such exemptions are known as 'permitted development'. A second statutory instrument, the Town and Country Planning (Use Classes) Order 1987, groups land uses into different categories and determines whether a 'material change of use' is occurring or not in respect of development proposals.

7.2.3.3   Planning application procedure
If the proposed development does not fall within the permitted development rights set out in the GPDO then a planning application may be required. The majority of planning applications are made to the relevant district authority. The main exemptions are applications for development which occur within a national park, or proposals relating to minerals or waste disposal. In the latter cases, the applications are directed to the relevant county planning authority, unless, of course, the proposed site is within a unitary authority, in which case it is dealt with by the district. The Secretary of State for DEFRA reserves the right to 'call in' any planning application for determination. This procedure usually occurs when a development proposal is particularly contentious. There are also occasions when the local authority itself is reluctant to make a potentially precedential decision and, therefore, refers the proposal to the DTLR for a decision.

The assessment of planning applications and the methodology for reaching a decision is significantly different in the United Kingdom from many other EU member states. Whilst the system is currently undergoing change and moving towards a more European model it is not yet correct to say that it is 'plan led'. Development control decisions rest on a number of criteria, amongst which the relevant plans are an increasingly important element, but the system continues to include reference to 'material considerations' outside of published plans. In this respect it allows for more flexibility, uncertainty and political influence than do many of its European counterparts. It is not reasonable to suggest that a development proposal must always accord with the plan in order to achieve consent, or that it will necessarily achieve consent if it does. In most cases, however, conformity with the plan is an important factor, and the DTLR is likely to become involved in exceptional cases. Having said that, it is significant that around

20% of applications for planning permission in the United Kingdom are unsuccessful whilst only around 5% in Spain and France are unsuccessful. This disparity reflects the use of 'other material considerations' in the United Kingdom as against strictly plan led systems in those other EU member states. It is also significant that there is no clear definition of 'material considerations'. This concept is continually developing and rests on a growing body of precedent and case law. Whilst government policy, as expressed in PPGs and the various types of plan, is a guide, under the current system it is not possible to predict with certainty whether a particular development proposal will be accepted or not.

It should be noted that to lodge a planning application ownership of the land or building intended for development is not obligatory. However, planning permission for a development bestows no legal right to carry out that work. In addition to which, the applicant is obliged to notify all parties with an interest in the property or land prior to any application. It is, therefore, possible to gain permission to carry out work on land and/or buildings prior to having a legal right over the property. For any proposed work which falls within the definition of 'development' and not within the 'permitted development' of the GPDO, a planning application is required. There are two types of planning application:
– Full Planning Permission, and
– Outline Planning Application.

Application for Full Planning Permission requires all the information salient to the development. The most important issues generally investigated by the local authority planners are siting, design, external appearance, means of access, floor space, numbers employed, felling of trees, drainage and landscaping. This allows the planner to evaluate the impact of the development accurately. With an application for Outline Planning Permission, the concept of the development is analysed and its acceptability in planning terms is assessed. The application need not detail the siting, design, external appearance, means of access or the landscaping. However, these 'reserved matters' will need to be approved at a later date. The advantage of this type of application over a Full Planning Application is that it requires a lot less time and money initially. In the case of land designated as 'sensitive' by Article 1(5) of the Town and Country Planning Act, Outline Planning Permission cannot be applied for.

7.2.3.4   Planning decisions and appeal system
The decision whether to give consent or refuse the application to develop is made by the Planning Committee. The members of the Committee are elected members of the local council. They make their decision based upon a

recommendation made by the planning department and presented in the form of an officers' report and public opinion for example, expressed in letters, directed to them concerning the development. Public consultation also plays an important part in the planning department's procedures.

Appeal against refusal of planning permission is made to the DTLR. Only the applicant or his agent may appeal. There is no third party appeal system such as exists in many other EU member states. Grounds for the appeal must be clearly stated and must relate to the original reasons for refusal. The DTLR decision will, as with the original decision, take into account the provisions of central government policy and local authority plans as well as any other 'material considerations'.

7.2.4   Environmental management

Clearly, the active and positive involvement and contributions of private companies are essential in implementation of environmental policy. Generally, UK private companies are beginning to take up the challenge of environmental protection against a background of voluntary rather than mandatory action on environmental management promoted by UK government. The British Standards Institution published BS 7750; as the first standard for a practically-applicable Environmental Management System (EMS) in the United Kingdom. This standard, which has now been superseded by ISO 14001 (British Standards Institution, 1996), is a voluntary certification system to protect the environment by quality management. So far, this standard, in either UK or international form, has been adopted by 15,000 companies in the United Kingdom. However, there is considerable concern about the slow pace of progress with environmental management throughout the UK economy at large. Many companies regard environmental concerns as a threat rather than an opportunity. In particular, small- to medium-sized enterprises (SMEs) have been slow to adopt environmental management practices. This is generally perceived as a result of lack of time, finance and capability. Various UK and EU schemes which provide subsidies for involvement in environmental management are being directed towards this problem. The reasons why this is an important issue is that SMEs are extremely numerous, they are connected to larger companies through the supply chain and they are regarded by some politicians as the main driving force for innovation and economic development.

One particular environmental policy instrument which is regarded as a fundamental means of promoting commitment to environmental management by private industry is the Eco-Management and Audit Scheme (EMAS). This scheme (Department of the Environment, 1995a) is intended as a useful adjunct to European Commission Regulation 1836/93 (European

Commission, 1993b). As an EU-wide action, EMAS *"provides clear guidelines and encourages managers to set and reach their own objectives"* (Department of the Environment, 1996a). Companies which are successful in their adoption of this scheme are awarded the EU-recognised EMAS symbol. In 1996, Mr. T. Garvey, Deputy Director General of DG XI of the European Commission, praised the then-Conservative government for encouraging the extension of EMAS to local authorities, thus providing a policy tool to assist the delivery of commitments to Local Agenda 21, and *"for going beyond normal implementation consistent with legal transposition of a community directive"* by developing a successful marketing campaign, which could be seen as an effective approach that other EU member states could adopt (Department of the Environment, 1996b). As such, EMAS does not provide any direct financial incentive for companies and other organisations which commit themselves to environmental management, although indirect financial benefits may arise through subsequent cost savings and favourable publicity.

However, some direct financial help is available for SMEs in the United Kingdom by means of the Small Company Environmental and Energy Management Assistance Scheme (SCEEMAS) which was launched in 1995 (Department of the Environment, 1995b). Originally, this scheme provided grants covering 50% of the costs of consultancy work involved in achieving EMAS registration by SMEs engaged in manufacturing, power generation, waste disposal and recycling. Subsequent rule changes resulted in the scheme applying only to manufacturing companies with less than 250 employees and an annual financial turnover lower than £32 million (€48 million). Additionally, the SCEEMAS grant was set at between 40% and 50% of consultancy costs, depending on the level of commitment to EMAS achieved by the company. These levels of increasing commitment include adoption of an environmental policy, preparation of an environmental review, development of an environmental programme, creation of an environmental management system, conducting an environmental audit, and, finally, as part of EMAS registration, publication of an environmental statement with subsequent validation procedures and submission of a formal application (Department of the Environment, 1996c).

It has been suggested that regulatory instruments of environmental policy encourage private companies to respond by establishing environmental management systems (Gouldson and Murphy, 1998). A number of private companies in the United Kingdom have, indeed, implemented such systems, either within the framework of ISO14001 and EMAS or otherwise. However, the motivations for such action are quite diverse and, in many instances, seem to relate as much to a desire to improve or extend their public image as a corporate response to regulatory pressure. The lack of mandatory requirements regarding environmental management systems is probably responsible for confusion over motivation on such action. Similar concerns arise from the voluntary approach to corporate environmental reporting. Despite government encouragement and appeals to private companies for independent auditing and benchmarking of environmental reports (Department of the Environment, Transport and the Regions, 1998a and 1998b) reporting standards are extremely diverse in the United Kingdom. However, government advice and assistance for private companies on environmental matters is improving through such services as the Environment and Energy Helpline. Additionally, an increasing number of private companies are adopting recognised standards for environmental management systems and are validating environmental reports with independent auditors. Hence, there have been incremental advances with environmental management within the essentially voluntary context of this activity in the United Kingdom.

7.2.5   Fiscal measures

In addition to direct environmental policy instruments, there are numerous other mechanisms which, intentionally or otherwise, promote environmental policy indirectly in the United Kingdom. The first fiscal measure declared as an obvious environmental tax in the United Kingdom was the landfill tax which was introduced in 1996. This tax was originally levied on the disposal of waste to landfill sites at a rate of £2 per tonne (€3.0) for inert waste and £7 per tonne (€10.5) for all other controlled waste. A subsequent increase of the landfill tax on these controlled wastes to £10 per tonne (€15.0) was announced by the Chancellor of the Exchequer in the budget in April 1996, with further increases of £1 per tonne (€1.50) each year until 2004 (Department of the Environment, Transport and the Regions, 1998c; Cowe, 1999). The effect of this tax is intended to shift waste disposal practices away from the use of landfill sites, which are regarded as environmentally less acceptable and a declining option, and towards incineration, in the first instance, and, ultimately, re-use and recycling. Apart from its direct economic effect on the costs of different waste disposal practices, the landfill tax has another benefit since revenues collected are channelled to projects which improve the local environment.

Other typical examples of indirect environmental policy instruments can be found in the field of energy. In particular, the implications of the Non-Fossil Fuel Obligation (NFFO) and the activities of the Energy Saving Trust (EST) are noteworthy. Both these instruments are potentially important in terms of reducing carbon dioxide emissions and could contribute significantly to meeting UK commitments under the Climate Change

Convention. However, it can be argued that this was not the original intention of either of these instruments. Both are, in effect, subsidy mechanisms. NFFO was originally devised as a means of providing financial support for the nuclear power industry in the United Kingdom, which was finally exposed as an uneconomic option during privatisation of the electricity supply industry. Although the vast majority of the subsidy (99% but declining), generated by a levy on all electricity bills, has supported the nuclear power industry, NFFO has fortuitously become the main means of nurturing the new renewable energy industry in the United Kingdom (Department of Trade and Industry, 1995). Despite its failings, NFFO has been instrumental in increasing the percentage of electricity produced from renewable energy sources which are abundant in the United Kingdom. During 2000, it was announced that NFFO support for renewable energy schemes would be replaced by a system of legally-binding targets, referred to as the Renewables Obligation, on regional electricity companies.

The EST is also funded through a levy on energy, via both electricity and natural gas bills. The activities of the EST promote improvements in energy efficiency and, thereby, result in environmental benefits, despite its prominent original justification of increasing the economic performance of UK industry. In general, the EST subsidises energy efficiency measures by means of grants which provide joint funding for specific projects selected through a competitive proposal process. Further impetus to environmentally-beneficial developments, such as renewable energy technologies and energy efficiency measures, respectively, is now being given by the introduction of an industrial energy tax, referred to as the Climate Change Levy, in the United Kingdom. The foundations of this development were prepared by Lord Marshall's review of possible economic instruments which would assist the UK government meet its targets for reducing carbon dioxide emissions (Lord Marshall, 1998). This new tax was applied in April 2001 to all industrial and commercial energy users, including agriculture and public administration but excluding transport. The tax has initially been levied at a rate of 1.17 pence per kg (€0.018) on coal, 0.96 pence per kg (€0.014) on liquefied petroleum gas, 0.15 pence per kWh (€0.002) on natural gas, and 0.43 pence per kWh (€0.006) on electricity. Revenue generated by this tax will be used to reduce national insurance (social costs) paid by employers, so that it can be regarded as a fiscal measure designed to support employment. Originally, the tax was expected to raise £1.75 billion (€2.63 billion) during its first year of operation (Cowe, 1999). Following consultations and lobbying by large industrial energy users, the Climate Change Levy was modified, chiefly in relation to exemptions for consumers who implement energy efficiency programmes.

This tax has been specifically presented as a means to assist UK commitments to the Climate Change Convention. In this context, the tax should reduce carbon dioxide emissions by 1.5 million tonnes during its first year of application. However, it should be noted that, despite this interpretation, the Climate Change Levy is not, as currently structured, a 'carbon tax' in the strict meaning of such a fiscal measure. The reason for this is that the rate at which it is proposed this tax will be levied is not related to the carbon content of the energy sources affected. Indeed, it was originally proposed that the blanket rate would be applied to electricity, in particular, so that there would be no distinction between electricity generated by fossil fuels, which produce carbon dioxide, and renewable energy technologies, which do not. Additionally, the tax is relatively small in comparison with other fiscal measures adopted for environmental purposes in countries such as Denmark and Sweden. However, this does not mean that the UK Climate Change Levy cannot be modified at a later date so that it does reflect carbon content. Furthermore, it has the potential to alter the competitive energy market in favour of renewable energy technologies and energy efficiency measures. As such the Climate Change Levy joins other fiscal measures, such as the landfill tax, which are gradually changing the economic climate for the promotion of new environmental technologies.

7.3   Environment-oriented technology policy

7.3.1   Historical context

In the United Kingdom, all elements of environment-oriented technology policy which may occur must be seen within the general context of technology policy. Historically, the simplest model of technology policy involves aims and objectives determined by the state, with actions funded principally through public finance and implemented chiefly by means of nationalised companies and similar organisations. This model is characterised as 'command-and-control' technology policy. Such a model was prominent in the United Kingdom during the 1950s, 1960s and 1970s. This somewhat traditional form of technology policy was generally extolled as a major potential force for modernising industry during the 1950s. It gained its full expression in the 'White Heat of the Technological Revolution' of Harold Wilson's Labour government of the 1960s. Increasingly larger amounts of public finance were directed towards substantial programmes of research and development (R&D) conducted by government departments, nationalised companies, and special agencies. However, even at the height of its public prominence, the 'White Heat' had begun to cool and, by the end of the 1970s, it had finally lost its popular appeal as the full extent of a number of technology policy failures became increasingly apparent. These failures included the technical and economic disasters of the Advanced Gas-cooled Reactor and Fast Breeder Reactor Programmes (Sweet, 1980; Bunyard, 1981) and the commercial debacle of the Concorde supersonic airliner project (Wiggs, 1971; Wilson, 1973).

In general, such 'command-and-control' technology policy was driven by an obsession with the engineering challenge and no real understanding or appreciation of commercial need. This type of approach was even applied to R&D on environment-oriented technologies, including renewable energy technologies. In late 1973 and early 1974, at the height of the first oil shock, when such technologies were being proposed as alternative energy options, it was decided that the Energy Technology Support Unit (ETSU) would oversee renewable energy R&D in the United Kingdom. Despite being a part of the UK Atomic Energy Authority (UKAEA), which was responsible for nuclear fission and fusion R&D, ETSU demonstrated real enthusiasm and serious commitment to R&D effort on a range of renewable energy technologies. However, policy imperatives seem to have resulted in over-emphasis on national solutions based large-scale engineering. Additionally, a lack of business experience, in general, and project management, in particular, was apparent. Although ETSU achieved some notable successes, the UK renewable energy technology R&D programme attracted considerable criticism from the House of Commons Committee of Public Accounts when it reported in July 1994 (House of Commons Committee of Public Accounts, 1994).

Specific criticism was directed towards the wind energy R&D programme. The Committee commented that:

"We are disappointed that, despite £54 million (€81 million, JG/NM) being spent on the wind energy programme, 84 per cent of installed wind turbine capacity in the United Kingdom is provided by imported manufacture. We note that the influence of the Central Electricity Generating Board (the nationalised company then responsible for supplying electricity for England and Wales) led to an emphasis within the wind energy programme on large scale machines; and that nearly £6 million (€9 million) was also spent on vertical axis machines. We note that neither of these technologies are currently viable" (House of Commons Committee of Public Accounts, 1994).

It has been pointed out that, in contrast, the success of the Danish wind turbine manufacturing industry, in both domestic and export markets, has been based on developing small-scale machines, incorporating established technology, and gradually increasing size as experience was gained in commercial production and operation (Elliott, 1997). As part of broader criticism of the renewable energy technology R&D programme, the Committee noted that:

> "...between 1975-76 and 1992-93 the Department's (of Trade and Industry, JG/NM) renewable energy research, development and demonstration programme had cost £340 million (€510 million, JG/NM)...", but considered that "...it is very doubtful that the relatively modest increases in new electricity generation justify the large sums spent..." (House of Commons Committee of Public Accounts, 1994).

Politically, these and other, more prominent failures were presented by successive Conservative governments as the obvious consequence of government interference in the legitimate freedom of private industry. Subsequent adherence to monetarist economic policy, combined with decreasing public expenditure and a laissez-faire approach to private industry, effectively marked the end of the traditional application of technology policy in the United Kingdom. During the 1980s, conventional political wisdom regarded any government policy which might interfere with the operation of the free market as totally unacceptable. Hence, a policy vacuum was created. Technology policy was increasingly left to private industry. Public R&D funding declined sharply and, over a period of time, those previous agents of government policy, the nationalised industries, were privatised. However, it was inevitable that an alternative would eventually come to fill this policy vacuum. By the 1990s, it was becoming clear that, left to its own devices, UK private industry would not invest in the R&D effort needed to innovate and compete effectively in world markets. Indeed, both the government and private companies were criticised for too much emphasis on short-term returns and too little commitment to long-term investment.

7.3.2   Technology policy instruments

7.3.2.1   The Foresight Programme
In 1993, partly in response to the apparent failings of the laissez-faire approach to R&D, the Conservative government launched the Foresight Programme as part of a major review of science, engineering and technology policy in the UK (Office of Science and Technology, 1993). By design and necessity, the Foresight Programme was based on the creation and maintenance of partnership between private companies, the research community, and government. The Foresight Programme is managed by the Office of Science and Technology (OST) within the Department of Trade and Industry (DTI). A key feature of the Foresight Programme is the

identification of R&D priorities which are expected to lead to innovation, competitive advantage, and wealth creation in the United Kingdom by private industry. In order to achieve this, sixteen Sectoral Panels, representing important aspects of the UK economy, were established during the first round of the Foresight Programme. Initially, environment-oriented technology policy chiefly fell within the scope of the Sectoral Panel on 'Natural Resources and Environment', although it should be appreciated that aspects of environment-oriented technology could feature in the coverage of any one of the other Sectoral Panels. Within this Sectoral Panel, the Sub-Panel on 'Cleaner Technologies and Processes' specifically addressed technologies, techniques, products and processes which reduce environmental impact and promote sustainable development.

Priorities were established within the first round of the Foresight Programme by reviewing relevant sectors of the economy and consulting with those within private companies, the research community, and government. This was achieved by means of networking, market analyses, scenario-building, sub-group discussions, regional workshops, and surveys, including a national Delphi survey. As a consequence, generic priorities were classified according to their attractiveness, interpreted as the perceived benefits from R&D and the ability of the UK economy to realise these benefits, and their feasibility, expressed as the expectation of tangible results and the ability of the UK scientific community to achieve such results (Office of Science and Technology, 1996). The specific environmental technology priorities identified by the 'Cleaner Technologies and Processes' Sub-Panel included separation technologies for waste management and water recycling, biotech processes for replacing high energy chemical and physical processing, cleaner technology for heavy engineering, cleaner coal technologies, and soil and ground water remediation (Office of Science and Technology, 1998b). Following dissemination of the preliminary findings, consultation began on the second round of the Foresight Programme (Office of Science and Technology, 1998a). In particular, comments were requested on the number and focus of the Sectoral Panels. Subsequent responses led to the numerous changes, including the merger of the 'Energy' Panel with the 'Natural Resources and Environment' Panel to form the new 'Energy and Environment' Panel (Office of Science and Technology, 1998c). This Sectoral Panel now chiefly encompasses relevant aspects of environment-oriented technology policy.

As intended, the outcomes of the Foresight Programme are developed and realised through partnerships between private companies, the research community and government. R&D funding is provided through government departments, the Research Councils (which are the established agencies for supporting basic, strategic, and applied research principally in the higher education sector) and by private industry itself. It has been argued that there is little new R&D funding available from the activities of the Foresight Programme since it chiefly derives from the re-naming and re-organisation of existing budgets. However, there is significant emphasis on matched funding which has enabled public funds to provide leverage for the release of private investment in R&D. In particular, collaborative research is supported by Foresight LINK Awards. The Awards offer up to 50% grant support for the cost of pre-competitive research projects between industry and universities or other parts of the research community. Emphasis is placed on research priorities which have been identified by the Foresight Programme, and SMEs are encouraged to become involved. The success of the Foresight Programme in the United Kingdom has led to it being promoted by the government as the new model for technology policy which could be adopted internationally, both in the European Union and elsewhere. Already, a number of countries have expressed interest in developing their own policies by incorporating the approach of the UK Foresight Programme.

7.3.2.2   Market promotion

Apart from financial assistance for pre-competitive research (see section 7.3.2.1) and support for the research, design and development of technologically-innovative products and processes (see section 7.4.2), there are no specific mechanisms for subsidising environment-oriented technology in the United Kingdom. However, there are at least two important initiatives, which can be interpreted as indirect policy instruments, for stimulating and promoting market demand for environment-oriented technologies produced by UK private industry. The first initiative is the Environmental Technology Best Practice Programme, which was launched in June 1994 as a five year joint action by the DTI and the former DoE. The purpose of this initiative is to raise awareness and to disseminate information on practical experience as a means of encouraging potential users to adopt best environmental technology for minimising waste and reducing environmental impacts. Relevant material is provided through the Environmental and Energy Helpline which is operated by ETSU. The second initiative is the Joint Environmental Markets Unit (JEMU), which is based in the Environment Directorate of the DTI. The aim of JEMU is to raise the profile of the UK environment-oriented technology industry within this country and abroad, with particular emphasis on increasing awareness of export opportunities. JEMU maintains a database of information on environmental markets in the UK and overseas, and on the capabilities of UK suppliers of environmental products and services. Additionally, JEMU administers the UK Technology Partnership which acts as a networking facility for environment-oriented technology transfer to developing countries.

7.4   Framework for innovation

7.4.1   Competitiveness policy

In the absence of an explicit, comprehensive and coherent environment-oriented technology policy in the United Kingdom, the broader framework which is intended to promote innovation generally assumes considerable importance. Indeed, recent successive governments have supported a policy aimed at creating suitable conditions within which innovation can prosper as a means of initiating and sustaining the competitiveness of the UK economy. This fundamental policy has been established and articulated in a series of White Papers prepared by the DTI and adopted by government. However, it was not until the third White Paper in 1996 that the impact of environmental action on competitiveness was specifically addressed, as in a separate chapter on the environment, it was recognised that *"firms which manage their environmental impacts are more competitive"* (Department of Trade and Industry, 1996). In particular, attention was drawn to the combined effect of market forces, public pressure, and the financial implications of environmental liabilities and comparative environmental performance on innovative and improved product and process design. The then-Conservative government promised a *"flexible regulatory framework which stimulated innovation"* (Department of Trade and Industry, 1996) but was unable to implement this before a Labour government was elected in 1997.

A new initiative on competitiveness policy came in the form of the fourth White Paper of 1998 in which the Labour government committed itself to economic competitiveness in global markets based on *"productivity, ability to produce innovative products and create high-value services"* (Department of Trade and Industry, 1998). The environmental dimension of this aim was encapsulated in the statement that

> *"All business in the UK, large and small, manufacturing and services, low- and high-tech, urban and rural, need to marshal their knowledge and skills to satisfy customers, exploit market opportunities, and meet society's aspirations for a better environment"* (Department of Trade and Industry, 1998).

In contrast to its characterisation by models of 'command-and-control' or 'subsidy', subsequent policy is portrayed in terms of a 'compact with business' in which the government has *"a key role in acting as a catalyst, investor and regulator to strengthen the supply-side of the economy"* (Department of Trade and Industry, 1998). The foundation of this policy is explained with reference to the perceived main driving forces of growth and innovation in the UK economy which are capabilities, collaboration and competition. Capabilities include all the diverse and essential skills required to promote and exploit innovation. Collaboration is seen as a necessary approach in relation to the complexity of the processes involved in producing and marketing modern products and services. Competition is presented as the only realistic basis for economic activity and growth.

7.4.2   Policy implementation and influences

It has been proposed that the policy for competitiveness will be implemented by a means of a collection of established and new instruments which are intended to address all aspects of the process of modernising the UK economy. Amongst these are a number of specific instruments which are relevant to the promotion of innovative environment-oriented technologies. In particular, these include launching the second round of the Foresight Programme (see section 7.3.2.1), establishing a new out-reach fund for interaction between universities and business, increasing the DTI's Innovation Budget, considering the case for extending the Small Firms Merit Award for Research and Technology (SMART) scheme, promoting innovation in SMEs through Business Link, and creating an Enterprise Fund.

The new round of the Foresight Programme continues the partnership between government, the research community and private companies in order to identify new market opportunities and address important themes, especially sustainable development. It was announced that £10 million (€ 15 million) would be available to support projects by means of the Foresight LINK Awards. A new out-reach fund was established, with a proposed eventual annual budget of around £20 million (€30 million), to reward universities for *"strategies and activities which enhance interaction with business to promote technology and knowledge transfer"* (Department of Trade and Industry, 1998). The DTI's Innovation Budget was to be increased by over 20% during the next three years to provide government funding for innovation of approximately £220 million (€330 million).

The SMART scheme is a very important instrument which is formulated to assist individuals and SMEs to research, design and develop technologically-innovative products and processes for the national benefit. Originally, the SMART scheme was one part of a collection of funding mechanisms which also included the SPUR and SPURplus schemes, and the innovation element of the former Regional Enterprise Grants. The new SMART scheme, which was introduced by the DTI in April 1997, combines all these earlier sources of funds. SMART grants are awarded on a competitive basis for technical and commercial feasibility studies of innovative technologies, and for the development, up to the pre-production prototype stage, of new products or processes which demonstrate significant advances. Feasibility studies are funded at a rate of 75% of total costs, up to a maximum of £45,000 (€67,500). SMEs which win SMART grants for development projects receive funding at a rate of 30% of total costs, up to a maximum of £133,760 (€200,000), or £401,280 (€600,000) for exceptional projects. There are specific rules on the eligibility of SMEs for receiving SMART grants which depend on the number of employees, and the annual turnover or the annual balance sheet total.

The SMART scheme can often be the starting point for any individual who wishes to develop an innovative idea into a commercial product or process. As such, assistance with feasibility studies is a crucial step which, if successful, leads to pre-production prototype development. At this stage, SME involvement is usually required both in terms of practical engineering expertise and ability, and financial investment. Business Link is seen as one very significant instrument which can bring aspiring innovators together with experienced local companies. Business Link is a network organisation which is run at a local level by a private sector-led partnership of Training and Enterprise Councils (TECs), Chambers of Commerce, Enterprise Agencies, local authorities, government, and other business support providers. The network offers a comprehensive range of business support opportunities, services, and information. In particular, Business Link provides advice from Innovation and Technology Counsellors and Design Counsellors. Additionally, there are connections with Business Innovation Centres (BICs), which have been established by the European Commission, and seven Innovation Relay Centres in the United Kingdom, which are part of an EC network for disseminating and exploiting the results of EC-funded R&D.

Apart from technical assistance, Business Link is intended to access local commercial advice which can sometimes be related to investment fund-raising, as an essential component of the exploitation of an innovative development. In recognition of the importance of such fund-raising, the Labour government announced its intention to create an Enterprise Fund which was expected to total approximately £150 million (€225 million) over the next three years. This is based on partnership between government, banks and the venture capital industry. Specific measures which have been proposed include a national venture capital fund to support high-tech businesses at a very early stage in their development, new regional venture capital funds probably administered through Regional Development Agencies to provide small-scale equity for businesses with growth potential, and more assistance for the financial development of SMEs using the Small Firms Loan Guarantee Scheme.

In the United Kingdom, developing policy on competitiveness and related instruments for implementation, combined with the work of the Foresight Programme, can be regarded as the current main means for encouraging innovation in all relevant fields, including environmental technology. This is wholly consistent with the approach followed by successive governments since the early 1990's. In general, recent governments have chosen to avoid direct involvement in the process of innovation. Rather, the responsibilities of the state are seen in terms of creating suitable economic and other conditions which foster innovation by the research community and, principally, private companies. The extent of government involvement in innovation may only extend to partnership. However, other policies may influence innovation indirectly, as in the case of UK environmental policy and its related instruments which provide a basic framework within which environmental technology innovation may be either promoted or constrained.

Regulatory aspects of environmental policy create pressure on private companies to modify their behaviour. Potential responses can range from organisational to technical change. In its basic form, technical change may involve the installation of 'end-of-pipe' technologies which are designed as control measures for wastes and pollutants. As such, these technologies are not normally regarded as innovatory. Instead, environmental technology innovation is usually interpreted as 'clean technologies' which avoid the initial production of waste products and subsequent pollutants. Clean technologies can include 'incremental innovations' which may incorporate marginal improvements or modifications to existing technical solutions. At a further extreme, clean technologies may consist of 'radical innovations' which are based on totally new technical solutions.

The effects of UK environmental policy on these different types of technical responses have been considered in relation to regulatory pressure, in the form of IPC, and voluntary action, as represented by EMAS (Gouldson and Murphy, 1998). It has been argued that the implementation of IPC in the United Kingdom has had both positive and negative effects on environmental technology innovation. The relevant analysis centres on the application of BPEO and emphasis on BATNEEC within the IPC process in the United Kingdom. Due to an implementation style based on negotiation between the regulator and the regulated company, positive encouragement for innovation is implied as a result of possible knowledge transfer. However, it would appear that this is more than counterbalanced by negative consequences. In particular, it is suggested that, despite flexibility and interaction, regulation is weakened through this process of negotiation and this, in turn, leads to uncertainty and inconsistency, which undermines the confidence of private companies in innovative solutions.

BATNEEC itself would seem to be a force for discouraging innovation, since it clearly favours 'available techniques'. Hence, private companies are unlikely to risk the development of clean technologies when 'end-of-pipe' technologies may satisfy regulatory requirements adequately. Hence, *"a consensual and conciliative enforcement style also characterises the implementation of IPC"* (Gouldson and Murphy, 1998) and this may not favour environmental technology innovation in the United Kingdom. Although IPC should promote integrated new technology approaches, the need to compromise with control or abatement 'end-of-pipe' technologies is recognised where necessary or expedient. However, informal negotiation, perceived as the basis for applying BATNEEC in the United Kingdom, is regarded as an obstacle to innovation.

> *"Clearly such a process has the potential to reduce the imperative for innovation in regulated companies if the demands of regulation are made less stringent. Additionally, because the outcomes of such a process of regulation are inherently unpredictable, it may also reduce the propensity for innovation by introducing uncertainties and inconsistencies into the implementation process" (Gouldson and Murphy, 1998).*

If the mandatory aspects of environmental policy are, on the whole, uninspiring for environmental technology innovation in the United Kingdom, then it is possibly expecting too much for voluntary actions, such as EMAS, to provide the necessary impetus. Indeed, it is implied that the main technical impact of EMAS is to promote investment in monitoring technology and process specific techniques. It has been concluded that *"at present EMS (environmental management system) standards appear to be having a limited direct impact on technological innovation"* (Gouldson and Murphy, 1998). Hence, advances with environmental technology innovation in the United Kingdom may rely more on the exploitation of market opportunities than on any direct influence from UK environmental policy. As a consequence, both current policy on competitiveness and the promotion of the Foresight Programme may be considered as more appropriate responses and adequate means of implementing the equivalent of an environment-oriented technology policy in the United Kingdom. However, the success of this approach will depend, crucially, on the entrepreneurial nature of UK private companies and their appreciation of the environmental demands of export markets. Whether they can achieve this without a sound basis of sales in home markets remains to be seen.

7.5   Conclusions

Environmental policy in the United Kingdom is well-developed, based on relevant historical and current legislation. This policy is regulated and policed by recently-rationalised, national organisations. Local authorities are involved in environmental policy through the established planning system. Both EMAS and SCEEMAS provide means by which private companies can be engaged in environmental policy. However, despite their success in raising the profile of environmental management, these schemes may only have a very indirect effect on innovation with environmental technology. Over a longer period, heightened awareness of the need for environmental improvement, resulting from serious commitment to environmental management, may expand the market for new environmental technologies, thereby encouraging innovation.

Likewise, relatively new fiscal measures, such as the landfill tax, and the Climate Change Levy, may also alter the economic climate in which decisions are made concerning innovative environmental technologies. Additionally, both NFFO and the EST have played an essential role in environmental policy, albeit indirectly, in the United Kingdom. However, it should be noted that they do not necessarily promote UK environment-oriented technology R&D. All projects benefiting from these schemes must have current 'near-commercial' prospects and support is not restricted to the products of UK industry specifically.

Environmental policy in the United Kingdom cannot be wholly characterised by the standard 'command-and-control' model. In general, environmental control in the United Kingdom tends to be regulatory rather than policy-led. The effect is a situation which penalises wrongdoers and offers little support to those which comply or even innovate. Hence, it can be argued that UK environmental policy does not actively promote environment-oriented technology policy. The application of IPC and the reliance on BATNEEC in the United Kingdom may also discourage environmental technology innovation in the form of clean technologies.

An explicit, comprehensive and coherent environment-oriented technology policy does not exist in the UK. A 'command-and-control' model of technology policy, in general, was applied in the past, between about 1950 and 1980. This was subsequently replaced by a 'laissez faire' approach which has been superseded by a 'compact with business'. This is set within the broader policy on competitiveness which is intended to provide suitable general conditions for promoting technological innovation. This involves government as a 'catalyst, investor and regulator' in partnership with private industry and the research community.

The prominent expression of this new approach to technology policy is the Foresight Programme. As such, the model represented by the Foresight Programme cannot be regarded as a standard 'subsidy' approach to technology policy. There may be indirect and/or unintentional benefits for environment-oriented technology R&D where subsidies apply in the United Kingdom. However, the provision of such subsidies is not a major policy instrument. Instead, the approach championed by the Foresight Programme is characterised by consultation, networking and partnership between key players in private companies, the research community and government.

Various policy instruments are available or are being introduced to support UK policy on competitiveness. Of particular relevance to innovation with environmental technologies are joint funding through the Foresight LINK Awards and the SMART scheme. Additionally, Business Link, as a local network organisation throughout the United Kingdom, is a very significant instrument for assisting both individuals and SMEs involved in the exploitation of technological innovation. Such support can provide not only information and advice but also contacts with experienced local companies which are necessary for the commercialisation of innovative technologies. Other relevant policy instruments recognise the need for access to investment which is crucial to practical commercial exploitation.

REFERENCES

British Standards Institution (1996) *Environmental Management Systems: Specification with Guidance for Use*. London.

Bunyard, P (1981) *Nuclear Britain*. London: New English Library.

Cowe, R. (1999) Green Light for Clampdown on the Polluter. *The Guardian*, 9 March, p.15. Manchester.

Department of the Environment (1990) *This Common Inheritance: Britain's Environmental Strategy*. London: HMSO.

Department of the Environment (1994) *Sustainable Development: The UK Strategy*. London: HMSO.

Department of the Environment (1995a) *European Commission Eco-Management and Audit Scheme: A Participan''s Guide*. London.

Department of the Environment (1995b) *The Small Company Environmental and Energy Management Assistance Scheme: Users Guide*. London.

Department of the Environment (1996a) EMAS Roadshow: Government and CBI in Partnership. *Energy Management*, January/February Issue: 7. London.

Department of the Environment (1996b) Europe Praises UK as First Local Authorities Register under EMAS, *Energy Management*, May/June Issue: 9. London.

Department of the Environment (1996c) EMAS: A Baker's Dozen, *Energy Management*, July/August Issue. London.

Department of the Environment, Transport and the Regions (1998a) *Environmental Reporting: Getting Started*. London.

Department of the Environment, Transport and the Regions (1998b) Minister Launches Bid to Raise Environmental Competence Standards, *Energy and Environmental Management*, September/October Issue: 13. London.

Department of the Environment, Transport and the Regions (1998c) *Less Waste More Value: Consultation Paper on the Waste Strategy for England and Wales.* London.

Department of the Environment, Transport and the Regions (2000) *The Air Quality Strategy for England, Scotland, Wales and Northern Ireland: Working Together for Clean Air*, Cm 4548, SE2000/3. London: The Stationery Office.

Department of Trade and Industry (1995) Information of the Non-Fossil Fuel Obligation for Generators of Electricity from Renewable Energy Sources. *Renewable Energy Bulletin*, No. 6. London.

Department of Trade and Industry (1996) *Competitiveness: Creating the Enterprise Centre of Europe.* London

Department of Trade and Industry (1998) *Our Competitive Future: Building the Knowledge Driven Economy*, Cm 4176. London.

Elliott, D (1997) *Renewables Past, Present and Future: The UK Renewable Energy Programme.* Milton Keynes: Energy and Environment Research Unit, The Open University.

European Commission (1993a) *Towards Sustainability: A European Community Programme of Policy and Action in Relation to the Environment and Sustainable Development.* Brussels: Office of Publications of the European Community.

European Commission (1993b) Council Regulation (EEC) No. 1836/93 *Official Journal of the European Communities.* Brussels.

European Commission (1996) Council Directive Concerning Integrated Pollution Prevention and Control 96/61/EC, *Official Journal of the European Communities.* Brussels.

Gouldson, A. and Murphy J. (1998) *Regulatory Realities: The Implementation and Impact of Industrial Environmental Regulation.* London: Earthscan Publications.

Grant, J. F. (1994) *Land Designations from a Planning Perspective Report No. SCP8/6.* Sheffield: Resources Research Unit, Sheffield Hallam University.

House of Commons Committee of Public Accounts (1994) *The Renewable Energy Research, Development and Demonstration Programme 42nd Report.* London: HMSO.

Lord Marshall (1998) *Economic Instruments and the Business Use of Energy.* London: HMSO.

Office of Science and Technology (1993) *Realising Our Potential.* London: Department of Trade and Industry.

Office of Science and Technology (1996) *Winning Through Foresight: A Strategy Taking the Foresight Programme into the Millennium.* London: Department of Trade and Industry.

Office of Science and Technology (1998a) *Consultation on the Next Round of the Foresight Programme.* London: Department of Trade and Industry.

Office of Science and Technology (1998b) *Sustainable Technologies for a Cleaner World.* London: Department of Trade and Industry.

Office of Science and Technology (1998c) *Blueprint for the Next Round of Foresight.* London: Department of Trade and Industry.

Sweet, C. (ed) (1980) *The Fast Breeder Reactor: Need? Cost? Risk?* London: Macmillan Press.

Wiggs, R. (1971) *Concorde; The Case Against Supersonic Transport, Friends of the Earth.* London: Ballantine.

Wilson, A. (1973) *The Concorde Fiasco.* Harmondsworth: Penguin Books.

World Commission on Environment and Development (1987) *Our Common Future.* Oxford: Oxford University Press.

Chapter 8

Synthesis

SABINE SEDLACEK
Department of Environmental Economics and Management, University of Economics and Business Administration Vienna, Austria

GEERTEN J.I. SCHRAMA
Center for Clean Technology and Environmental Policy, University of Twente, the Netherlands

8.1   Introduction

The general focus of this book is on national policies and their role in stimulating environment-oriented innovation. Within the ENVINNO research project, we have focused on technological innovation and defined environment-oriented technologies in a manner that implies that the amount of production residuals can be reduced or the quality can be changed, so that they present less of a hazard to the environment. Beside market incentives, environmental policy and technology policy are the main public drivers stimulating companies' innovation activities. The creation of 'green markets' therefore, partly depends on such public stimuli which can provide the basis for green businesses. Hence, the aim of the policy analysis was to determine the effectiveness of these two policy fields.

The book includes six national reports that present an analysis of these two policy fields, i.e. environmental and technology policy. The reports are structured to reflect the development of both policy fields during the last three decades, their institutional framework, the most significant instruments implemented and finally the co-operation efforts (policy approach). As defined in the chapters before each of the six reports presents an individual national analysis, based on the specific institutional and general policy

Geerten J.I. Schrama and Sabine Sedlacek (eds.) Environmental and Technology Policy in Europe. Technological innovation and policy integration, 225-240. © 2003 Kluwer Academic Publishers. Printed in the Netherlands.

framework. Therefore, the resulting product is not a comparative study, but an attempt to synthesise elements of six national policies in order to shed light on a European framework of environment-oriented innovation stimulation. The following two sections present the synthesised trends and findings and not a conclusion coming of the six national studies.

8.2   Policy approaches

Research on innovation-oriented environmental policy and its regulation mechanisms has been carried out successfully all over Europe during the last few years, e.g. the research programme of the German consortium on 'Innovation Impacts of Environmental Policy' (FIU), 1996-1998 (Lehr and Löbbe, 2000: 112f).[1] The focus of most of this research has been on specific policy instruments or tools and their impacts on innovation activities. The ENVINNO national studies presented in the chapters before are based on a different strategy, which defines environmental policy and environment-oriented technology policy approaches in a systematic manner. Policy instruments are elements of these approaches, but they are only one part of the regulation impact chain that consists of *"the institutional context, the actor constellation, and policy learning in networks and negotiation systems"* (Jänicke, 1996). Therefore, an analysis of policy approaches needs to cover all these elements. For the specific ENVINNO research focus on an environment-oriented, innovation friendly policy sphere in particular, both environmental and environment-oriented technology policy approaches need to be analysed in this manner. Such an environment-oriented, innovation friendly policy approach can be described as follows (Jänicke et al., 2000: 135), the policy should be:
–   based on dialogue and consensus,
–   calculable, reliable and have continuity,
–   decisive, proactive and ambitious,
–   open and flexible with respect to individual cases,
–   management and knowledge oriented.

As mentioned above, policy approaches consist of various elements. Jänicke et al. (1999 and 2000) define policy approaches as an integrative patchwork of three elements:
1. *Policy style*. To define a policy approach, a general style has first to be identified. The policy style is normally shaped by: the general management orientation, the flexibility for adaptations, a certain

---

[1]   For a detailed overview see Hemmelskamp, Rennings and Leone (eds.) 2000.

dissemination level, the general judgement and an orientation along dialogues and consensus. Altogether these characteristics inform us about the general orientation of a specific policy.

2. *Instrumentation.* Here, research has to concentrate on the cause – effect chains of various instruments. Firstly, this element includes the variety of instruments and their potential for economic stimulation. Secondly, a strategic or an isolated orientation determines the cause – effects of specific instruments. Finally, the question of an active support of innovation processes in their different phases is raised for discussion. The instrumentation to a more flexible and individual one.

3. *Actors' configuration.* All the above mentioned elements depend on the degree of involvement of relevant stakeholders. The type and intensity of such an involvement provides information about the general characteristics of a certain policy style. The variety of instruments and their operation is directly connected with the actors' configuration. Distinctive features are: the role of networks, the interaction between regulators and involved target groups and of course the interaction between various involved actors. In an integrative policy approach a higher degree of networking will foster policy integration.

To summarise, the policy approaches of the countries under study need to be analysed by examining these three elements. We finally identified four different approaches on the basis of the main ENVINNO results and findings. The six ENVINNO countries (Austria, Germany, Denmark, Spain, the Netherlands, and the United Kingdom) are characterised by different policy approaches resulting from their specific institutional and organisational landscape. As it seems, the approaches differ as far as the three elements, policy style, instrumentation and the actors configuration are concerned, and co-existed in the countries under study at different stages in time:

1. *Command-and-control approach.* This approach is top-down oriented; it implies a high level of government intervention and can be described as mandatory regulation. The policy objectives within this approach are facilitated in the form of environmental standards, frequently expressed in maximum emission limits for specific equipment. Policy in general is predominantly directed at industry and environmental regulation is embedded into the general system of permits and controls for companies. Environmental policy is relatively focussed on specific instrumentation. Goals tend to be flexible.

2. *Strategic approach.* With a significantly high level of government intervention (top-down orientation), then strategic approach is comparable to the command-and-control approach. One distinguishing

feature is that policy oriented along strategies, e.g. the national environmental plans in several OECD countries or the EU sustainable strategy. These strategies include specific targets and a wide range of different instruments to fulfil these targets within a certain period, in most cases for short or medium term periods and rarely for long-term perspectives.

3. *Target group approach.* The target group approach consists of voluntary rather than mandatory regulation (bottom-up orientation). Target groups are the main addressees for environmental regulation and are highly aware of the full scope of potential regulation mechanisms. They are actively involved in the regulation process, i.e. negotiations between policy actors and experts in different sectoral and spatial levels. In this case target groups are empowered actors in the policy network.

4. *Management oriented approach.* This is the first attempt to move towards an integrative top-down and bottom-up approach. Here, public administration is reorganised and fully integrated within the whole process. Stakeholder participation constitutes the most important progress. One elementary part of this participation process is a step towards inter-policy co-operation with the aim of defining long-term strategies and stringent goals. New instruments are strategic plans, worked out in co-operation with relevant stakeholders. The introduction of such incentive based policy instruments, does however, have a slightly different ring compared to developments discussed earlier. Conceptually, they are now seen as strategic options in an integrated environmental management system, striving for more efficiency, less bureaucracy and more self-interest guided steering principles.

Having now a picture of policy approaches in mind the dynamics of such approaches seem the most important factor for analysing different national approaches. Our main prepositions (Jänicke et al., 1999 and 2000; Faucheux and O'Connor, 1998; Faucheux, 2000) can be formulated as follows:

– Policy approaches are changing during a certain time period. The changes from a certain approach to another occur continuously, as a result of dynamic processes in a policy field.

– In almost every European country a top-down oriented approach is somehow the starting point for a newly developed policy field, e.g. environmental policy. In many cases a command-and-control approach is implemented in the first run. Environmental regulation was originally media oriented with the aim to solve problems of water, air or soil pollution.

– The weakness of top-down approaches is, on the one hand, a missing relation to the implementation environment policy (for instance missing feedback loop between regulator and addressees of regulations), on the other hand it is clearly limited in solving rather complex environmental problems.

– The consequence is a paradigm shift in order to develop long-term strategies that are able to cope with complex interdependencies. Such new paradigms have a strong need for goal orientation and participatory style, rather than the classical mandatory regulation. As mentioned above, there are three feasible policy approaches, i.e. strategic, target group oriented and management-oriented which occur within a given time or even coincide (see figure 8.1).

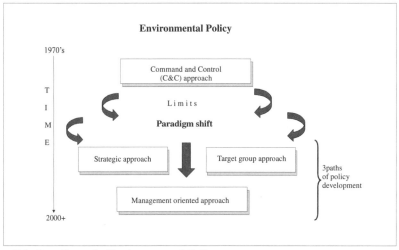

Figure 8.1: Paradigms in Environmental Policy

The approaches presented above can be identified in the legal framework of environmental and environment-oriented technological regulation and have clearly undergone some changes in approach and philosophy during the last three decades.

The early period of environmental legislation in this field was characterised by incident driven legislative activities. In the 1970s, a change could be observed, in some countries, towards a more systematic way of regulating environmental matters. A mainly command-and-control oriented approach was chosen – by which the residual receiving media (air, water, soil) were treated individually. Air pollution was tackled first[2] – early, specific regulations go back to the 19th century and even before. Water

---

[2] The Netherlands in 1971; United Kingdom in 1972; Spain in 1972; in Germany a series of laws were passed in the 1960s; Austria in 1980 and 1988; Denmark in 1974.

pollution was legally addressed at about the same time.[3] The issue of safe disposal of toxic waste caught the lawmakers' attention later,[4] followed by tackling problems of contaminated soil, still a very important issue today, usually linked with a debate on stricter liability rules.[5] Solid waste management has a long history of regulation attempts. Germany and the Netherlands were pioneers on legislative efforts for dealing with industrial waste, in the 1970s, while more general management issues dominated this sector in the nineties (separation, incineration, etc.).

Table 8 1: Media oriented regulation in the countries under study (some examples)

	1970s	1980s	1990s	2000 -
Water	Netherlands (1970)	Spain (1985)		
	Germany (1976)	United Kingdom (1985)		
Air	Netherlands (1971)	Austria (1980)		
	United Kingdom (1972)	Austria (1988)		
	Spain (1972)			
	Denmark (1974)			
Toxins	Netherlands (1976)	Germany (1980)		
		Austria (1987)		
Solid Waste	Germany (1972)		Austria (1990)	
			Germany (1991)	
			Germany (1994)	
			United Kingdom (1996)	
			Spain (1998)	
General Management	Germany (1972)		Germany (1990)	
	Netherlands (1979)		Netherlands (1993)	
			Germany (1998)	

The traditional command-and-control approach with its repairing orientation (i.e. ex-post regulation with emphasis on end-of-pipe technologies) was often criticised because of its neglect of natural links between the receiving media and the complex character of ecological systems (lack of systemic view). Early attempts to overcome this legal reductionism and to introduce more encompassing legal frameworks for resource management were

---

[3]   The Netherlands in 1970; Germany in 1976; in Spain and the United Kingdom more general regulations date back to the mid 1980s.

[4]   For example: the Netherlands in 1976; Germany in 1980; and Austria in 1987.

[5]   For example in Germany.

already started in the 1970s,[6] but are still an important issue today. This rethinking process taking place in many European countries, who were considering more systemic approaches, also stimulated the discussion on new management approaches replacing the well established, sector based, control and command approach of policy making (see figure 8.1).

In the 1990s, a paradigm shift towards national environmental plans (often including specific actions) and environmental programmes is becoming discernible, which can be indicated as a shift to the strategic approach. These action programmes are becoming more and more an integral part of an environmental management system based on clearly defined goals, usually laid down in national plans for the environment[7] or integrated, compatible action plans.[8]

In the process of integration and the establishment of compatible action plans, *technology policy* is also increasingly seen from this perspective. As environment-oriented technology policy, as pointed out in almost all six country reports, did not really exist in a systematic way until the 1990s: only a few legal foundations were necessary to make activities feasible, most of the R&D programmes were not earmarked for the environment and could easily be accommodated in the existing legal and institutional framework.

With given national plans to manage environmental resources more effectively, targeted technology action programmes were developed to assist environmental improvement efforts.[9] Increasingly the strategic elements of the various action plans and programmes were emphasised. The importance of newly developed and tested environment related technologies as a vehicle to enhance the economic performance of a country and to open up new market niches were recognised from 1992 onwards e.g. this idea is expounded in the documents starting a new export initiative of the Danish industry; in the UK, a new programme for 'best practice in environmental technology' was introduced in 1994, emphasising the marketing aspect.

While the 1990s were mainly characterised by (some kind of) strategic approach, the 1970s and eighties were mainly characterised by command-and-control and changes to a more target group oriented approach.

In the early phases of the command-and-control approach with its isolated focus on environmental media – water, air, waste, soil – specific instruments especially in the EP field were introduced. These instruments were directed towards specific media problems, e.g. a water levy. Their main

---

[6] For example in Germany in 1972, and the Netherlands in 1979.
[7] For example in the Netherlands (NEPP 1989); the United Kingdom (This Common Inheritance 1990); and Austria (NUP 1995).
[8] Denmark (1987-1991) and Spain (1995-2000).
[9] For example in Denmark ('Technology Push' 1985-1991); the United Kingdom (Foresight-Program 1993); and Austria (Technology Concepts 1989, 1996).

thrust was to remedy damages due to past pollution, or to reduce emissions at the source ex post, thus favouring end-of-pipe technologies and reuse and recycling options (repairing strategies).

The current policy analysis literature maintains that specific instruments cannot produce consistent policy results without explicitly considering systemic effects (Heritier, 1993; Jänicke and Weidner, 1995; Jänicke et al., 1999). Empirical analyses, taking the systemic perspective into account, including this report, emphasise the importance of dynamic interactions between public and private actors concentrating on strategic policy approaches (Jänicke and Weidner, 1995).

Policy approaches have changed their organisational pattern towards – 'management by objectives and results' – a development called 'New Public Management' in the policy analysis literature (Jänicke et al., 1999). This new approach advocates the formation of policy networks (management oriented approach). Additionally, a prevention oriented environmental policy and environment-oriented technology policy based on integrated technologies ('clean technologies') emphasises the chain of technological strategies to reduce pollution (transformation – emission – diffusion – residuals concentration) by focussing on the beginning of the process rather than the later stages as seen hitherto.

In the countries under study, we found the paradigm shifts indicated above (see figure 8.1) within different periods of time. Table 8.2 summarises these different paths which occurred step-by-step or even coincided.

Table 8.2: Policy paradigm shifts in the countries under study

Country	1970s	1980s	1990s	2000 +
Austria		command & control →	strategic and → target group	management oriented
Germany		command & control	target group →	management oriented
Spain		command & control →	target group →	management oriented
United Kingdom		mandatory regulation	strategic market driven: target group	
Denmark		Strategic →	target group →	management oriented
The Netherlands		command & control →	strategic and → target group	management oriented

Let us next summarise these developments, in more detail, for the participating countries:

Austria has a long tradition of the command-and-control approach. In the 1970s environmental policy almost exclusively acted with a strong media orientation, starting with regulation on air pollution, followed by water pollution in the 1980s. Technology policy was somehow complementing this direction with the establishment of support systems for specific technologies. These R&D stimulation programmes mainly focused on end-of-pipe technologies during the 1970s and on reuse and recycling technologies during the 1980s (repairing strategies). A tremendous paradigm shift occurred in technology policy in the 1990s: from R&D stimulation programmes towards stimulation of applications and wide spread diffusion. In the mid nineties, an environment–oriented technology policy came up with specific earmarked programmes (prevention orientation). The first strategic, and therefore, more systemic approach in environmental policy was implemented in the national environmental plan (NUP) in 1995. This was also the first attempt to implement the target group approach. Recent developments are fostering policy integration or co-operation efforts, e.g. the specific earmarked 'Program to Promote Sustainable Economic Activities' – a first attempt at a new management orientation.

In *Germany*, we also find a long tradition of the command-and-control approach, with a media orientation in the 1970s and eighties (repairing strategies). In the 1980s, informal agreements with industries formed corporate experts networks (target group approach). In the 1990s, the prevention principle has replaced repairing strategies. At the end of the 1990s, new information and negotiation tools were developed which was the starting point for the new management oriented approach. Recently, a more comprehensive perspective of environmental policy towards sustainable development is evident. Technology policy followed development paths comparable to the Austrian technology policy.

The command-and-control approach had a comparable importance in *Spain,* as in both countries discussed before. The media orientation in the 1970s and 1980s developed from repairing strategies against air pollution, to water and waste. The 1990s were significantly influenced by European Union directives requiring a new approach – the target group strategy. New market oriented instrument like tax allowances or sectoral agreements, Eco-labelling or EMAS were put into operation. The R&D support on the one side started relatively and comparably late, in the mid 1980s. Technology support, with an explicit environment orientation, started very early in the 1990s. This explicit, environment-oriented, technology policy now provides incentives for environmental policy to move toward a new management oriented approach.

In the *United Kingdom* we find a different environmental policy system which has a long tradition, but no explicit command-and-control approach. The 1970s with its repairing strategies were mainly dominated by mandatory regulation, which was completely abandoned in the 1980s and nineties. A new market driven approach occurred with a comparably low level of government intervention (voluntary rather than mandatory regulation) – the target group approach. It was based on information and negotiations with experts at the local level. The United Kingdom does not have an explicit technology policy- these tasks are considered part of the competitiveness policy.

Denmark is traditionally dominated by a consensus oriented and co-operative policy style which is contradictory to the command-and-control approach. Therefore, a significant degree of the implementation and administration of environmental protection tasks has been delegated to local authorities, since the early 1970s. The period from the 1970s to the mid eighties, was the starting point of the strategic approach, with guidelines and practices for a hierarchical set of ambient environmental quality monitoring and standards. From the mid-1980s onwards, environmental action plans were implemented in co-operation with all stakeholders (target group approach). This was also the starting point for an environment-oriented technology policy with specific clean technology programmes. In 1992, a paradigm change has taken place in environmental policy. Supply and demand side policies became responsible for fostering pollution prevention and technological innovation. A number of new instruments and specific action plans and programmes were developed in co-operation between environmental and technology policy, which can be indicated as a shift from the traditional strategic to the management oriented approach.

In the *Netherlands,* the command-and-control system was valid during the 1970s and eighties. In 1989, the first national environment plan (NEPP) came into action and this was the official break with the command-and-control approach. A more flexible and consensual approach appeared – the start of the target group principle. Technology policy programmes, with an explicit environment orientation, started in the 1980s. This period was rather emission oriented. An integrated focus on clean technologies appeared at the same time as environmental policy shifted to the target group approach. In the 1990s, a co-operative paradigm with partnerships involving technology developers and other stakeholders started, which could be seen as a step toward the new management oriented approach.

8.2.1   The new management oriented approach, a step towards policy integration

As pointed out in the section above, environmental regulation evolved from a policy approach with a high level of government intervention (mainly command-and-control), to a more integrative and voluntary regulation (management oriented approach). This new, management oriented approach is a step towards policy integration *"that would allow environmental objectives to be integrated into non-environmental policy areas"* (Gouldson and Murphy, 1998: 12). The involvement of different governmental and non-governmental actors and stakeholders requests specific co-operative and consensual tools. The quest for strategic options in an integrated environmental management system, striving for more efficiency, less bureaucracy and more self-interest guided steering principles needs to be discussed in detail. Policy integration, therefore, warrants co-operative and consensual strategies. These strategies demand co-operation models that are able to cope with these integration efforts. There are several models which co-exist (Schuh and Sedlacek, 2000 and 2002):

Traditional formal co-operation
These co-operations are established via a specifically created institution or corporation, founded due to of a need to co-ordinate two or more public authorities in a specific policy field or in a specific problem solution process. Typical examples of such co-operations are to be found e.g. in the form of planning institutions or traffic planning systems.

Formal co-operations without institutionalisation
These co-operations are based upon legal agreements, without founding specific institutions or corporations, which might carry out the tasks in question. The partners to these agreements can either be public authorities and/or private corporations. The most commonly known example of this type of co-operation is the public-private-partnership, which is found in many cases of public service provision (waste management, water treatment). This model is particularly project oriented and aims at cost sharing models.

Modern co-operation models following the theoretical concepts of management science
– *New public management* is often defined as a strategic approach in the field of environmental policy (Jänicke et al., 1999). It emphasises the

importance of defining specific targets, which need to be fulfilled with flexible instruments. Furthermore, it implements a consensual legitimisation ('stakeholder approach') by involving all the relevant actors.

- *Citizen participation* in public projects, especially large scaled projects: citizens organise themselves in participation groups with the aim of strengthen their position as single actors.
- *Mediation.* For large scaled public projects a mediation process can co-ordinate different parties in the implementation process. Especially for environmental impact assessments, a third, independent actor functions as a co-ordinator.
- The *Harvard model* is mainly applied in the United States and the United Kingdom. It is a resolution model with win-win character mainly based on arguing, which depends on complete information and the know-how of each actor. Various components often do not exist in the case of public projects: equal decision power, willingness to solve the problem, equal basis of negotiation and social competence.

Policy co-operation, as one form of structuring negotiation processes within and between policy fields, should be embedded in participatory structures, to cope with conflicts resulting from different policy interests. Such participatory structures can be found in almost every modern type of co-operation (see above). Therefore, policy integration should be moderated by external actors who are not part of the involved policy fields.

8.2.1.1 Driving forces for policy integration

Having a concept of possible co-operation strategies in mind, the question of stimulation of such integration processes arises. What are the driving forces which reinforce such integration efforts? As pointed out in the introductory chapter, EU environmental policy identified the necessity of integrating environmental requirements into other areas of policy (Third Environmental Action Programme in 1982). What kind of driving forces are stimulating concrete integration efforts? Recently, two important driving forces have been identified in connection with finding an effective policy to initiate and achieve sustainable development processes:

1. *A global problem solution framework as a driving force for policy co-operation.* A global problem solution e.g. the Kyoto Protocol can function as an impetus for policy co-operation. Global problems like the climate change challenge affect more than one policy field. The cause-effect cycle is (a) complex and (b) needs to be analysed within several policy fields. Examples from the ENVINNO sample can be found

in almost every country under study. One elaborated case is the co-ordination of the Dutch Ministry of Environment with the Ministry of Economic Affairs within the frame of the climate change debate. A second example, with concrete co-operative actions between the environmental policy and the technology policy areas, was given in the chapter on Denmark. Both ministries have worked out new R&D programmes, focusing for example on cleaner technology, to promote environmental management.

2. *Medium and long-term problem solutions.* There is a need for medium and long-term policy programmes in order to find adequate solutions for the complex problems mentioned above. The fact is that this kind of policy programme should fall under the responsibility of the potentially involved policy actors. These should formulate commonly defined goals based on commonly recognised trends and strategies. One of the most important factors for success to consider is a continuous up-date of commonly defined goals. Within the ENVINNO project, a prominent example for such a long-term oriented policy program influenced by innovation policy and environmental policy can be found in Spain the 'National R&D Program on Environment and Natural Resources'. The climate change challenge (see above) can certainly be identified as driving this effort., Several environmental issues, such as water quality, waste management, climate change, etc., which pertain typically to responsibility of several sectors, are dealt within this programme .

8.2.1.2   Types of policy integration by function
The requirement to integrate environmental issues into several policy fields was intensified in response to the above identified driving forces. As pointed out in the examples above, different functions are warranted to support an integration process. With respect to policy integration, three functions seem to be of importance: (1) organisational functions, (2) process oriented functions, and (3) output oriented functions.

1. Organisational functions
The adoption of organisational functions, embedded in an institutional framework, is a precondition to foster integration processes. The relevant ministries need more institutional background for their co-operation efforts. There is a strong need for organisations or institutions at the interface of the involved policy fields. These organisations function like moderators or even mediators between the stimulus sender (policy) and the receiver (e.g. industries). One elementary reason for such a mediation function is the necessity to control the observance of the rules for the above described inter-policy co-operation. Furthermore, these institutions should report milestones such as the degree of goal attainment etc.

In the ENVINNO sample we found several examples for such institutions, but they all differ with respect to organisational structures and orientations. In the Netherlands, the institutions Infomil and Syntens are organisations at the interface of environmental and technology policy. In the United Kingdom, the Office of Science and Technology (OST) for example manages the Foresight Programme. This Office established 16 sectoral panels with the aim of developing partnerships between private companies, the research community and government- which is indeed a question of moderation of all the groups.

2. Process oriented functions

Process oriented functions mainly address communication and information. Participatory mechanisms, for instance negotiations with target groups, need to cope with possible conflicts. There is a strong need to harmonise targets. The stakeholders involved are normally locked in their specific organisational settings. The establishment of intermediary stakeholders e.g. task forces, supports harmonisation.

Task forces. Inter-policy co-operation is an open process which needs to be communicated to all stakeholders – i.e. industries, research institutions, etc. For that reason, specific task forces are needed to communicate the previously developed and agreed contents of such inter-policy programmes.

In the Netherlands, such task forces are on duty to inform industries about the future needs in the field of new technologies on the basis of the technology programme. ('Programma Milieu & Technologie').

In Spain, such task forces were established in the late 1990s for the implementation of the IPPC directive, with the aim of informing industries about the scope of potential technological improvement.

In Austria, consulting firms often function as moderators for specific programmes, e.g. the Eco-Profit Programme.

3. Output oriented functions

Functions that aim at the output of integration processes are mainly product oriented. In the ENVINNO context, we identified three different products – co-operative policy initiatives, commonly shared instruments and inter-policy programmes.

Co-operative policy initiatives. Referring back to the above discussed medium and long-term orientation of policy programmes, the question of a co-operative and integrative formulation process arises. What are the strengths and weaknesses resulting from, e.g. an inter-ministerial co-operation? What are the critical factors?

ENVINNO examples of concrete integration between two or more policy resorts again come from the Netherlands, where four ministries –

environment, economic affairs, transport and agriculture have jointly developed a policy document 'Environment and Economy'. The problems involved in working together and finding a consensual strategy based on four different perspectives were manifold, but the product finally lead to a series of specific policy initiatives.

Commonly shared instruments. Within a next step, instrumentation needs to be discussed. If two or more policy fields co-operate in terms of developing common policy documents, they need to work out goals and strategies first. To fulfil such goals explicit instruments are warranted in the long run. Therefore, the question of commonly shared instruments is essential in terms of policy co-ordination. There are two possible ways which co-exist – the case of one jointly defined goal and different reactions or incentives coming from each policy field or the case of jointly defined goals and also common instruments. The instruments used are frequently voluntary agreements.

Inter-policy programmes. A much more intensive form of co-operation between two or more policy fields are inter-policy programmes which cover more than one perspective and overall strategy. This is the first step towards an intensive policy integration process. If several policy fields work out such programmes, they first need an agreed long-term perspective, e.g. a long-term perspective for sustainable development. Each policy field formulates specific incentives. Each incentive is not fully effective when isolated and not fully harmonised with other incentives, coming from other related policy fields. Different, isolated incentives address different target groups - one reason for their not being fully effective. Some types of incentives are diffusion-oriented, others pioneer-oriented. In many cases the different signals are not interconnected and are therefore competing. Without harmonising these signals or incentives an effective innovation support system will not be designed. Therefore, jointly defined and elaborated policy programmes are essential for successful and tailor-made innovation support systems.

The Danish case shows a high intensity of inter-policy programmes. In the late 1980s, The Ministry of the Environment and The Ministry of Trade and Industry formulated a common strategy and administered co-operative programmes to develop new technologies and implement management systems.

In the United Kingdom, the Foresight Programme is not a real example for an inter-policy programme although it was launched as a part of a major review of science, engineering and technology policy in 1993.

The Austrian Programme on Sustainable Economic Activities is a kind of inter-policy initiative between environmental policy and technology policy, although the programme was launched by the Ministry of Economic Affairs individually. Furthermore, the Eco-Profit Programme is an inter-policy programme where environmental, technology and economic policy goals are covered.

REFERENCES

Faucheux, S. (2000) Environmental policy and technological change: Towards deliberative governance. In: J. Hemmelskamp, K. Rennings and F. Leone (eds.) *Innovation-oriented environmental regulation. Theoretical approaches and empirical analysis.* ZWE economic studies 10. Heidelberg: Physica Verlag.

Faucheux, S. and O'Connor, M. (eds.) (1998) *Valuation for sustainable development. Methods and policy indicators. Advances in Ecological Economics.* Cheltenham, Northampton: Edward Elgar.

Gouldson, A. and Murphy, J. (1998) *Regulatory realities. The implementation and impact of industrial environmental regulation.* London: Earthscan Publications.

Heritier, A. (Hrsg.) (1993) *Policy-Analyse. Kritik und Neuorientierung.* Opladen.

Jänicke, M., Weidner, H. (eds.) (1995) *Successful Environmental policy. A Critical Evaluation of 24 Cases.* Berlin: edition sigma.

Jänicke, M. (ed.) (1996) *Umweltpolitik der Industrieländer. Entwicklung - Bilanz - Erfolgsbedingungen.* Berlin: edition sigma.

Jänicke, M., Kunig, Ph., Stitzel, M. (1999) *Umweltpolitik. Politik, Recht und Management des Umweltschutzes in Staat und Unternehmen.* Bonn: Dietz Verlag.

Jänicke, M., Blazejczak, J., Edler, D. and Hemmelskamp, J. (2000) Environmental policy and innovation: an international comparison of policy frameworks and innovation effects. In: J. Hemmelskamp, K. Rennings and F. Leone (eds.) *Innovation-oriented environmental regulation. Theoretical approaches and empirical analysis. ZWE economic studies 10.* Heidelberg: Physica Verlag.

Lehr, U. and Löbbe, K. (2000) The joint project 'Innovation impacts of environmental policy'. In: J. Hemmelskamp, K. Rennings and F. Leone (eds.) *Innovation-oriented environmental regulation. Theoretical approaches and empirical analysis.* ZWE economic studies 10. Heidelberg: Physica Verlag, pp. 109-123.

Schuh, B. and Sedlacek, S. (2000) City, Hinterlands – Sustainable Relations. *Proceedings of the 40th Congress of the European Regional Association (ERSA)* Barcelona, August 30th to September 2nd 2000.

Schuh, B. and Sedlacek, S. (2002): Problem Centred City-Hinterland Management – A Scientific and Policy Approach. *Proceedings of the 42nd Congress of the European Regional Association (ERSA)* Dortmund, August 27th to 31st 2002.

ENVIRONMENT & POLICY

1. Dutch Committee for Long-Term Environmental Policy: *The Environment: Towards a Sustainable Future.* 1994   ISBN 0-7923-2655-5; Pb 0-7923-2656-3
2. O. Kuik, P. Peters and N. Schrijver (eds.): *Joint Implementation to Curb Climate Change. Legal and Economic Aspects.* 1994   ISBN 0-7923-2825-6
3. C.J. Jepma (ed.): *The Feasibility of Joint Implementation.* 1995
         ISBN 0-7923-3426-4
4. F.J. Dietz, H.R.J. Vollebergh and J.L. de Vries (eds.): *Environment, Incentives and the Common Market.* 1995   ISBN 0-7923-3602-X
5. J.F.Th. Schoute, P.A. Finke, F.R. Veeneklaas and H.P. Wolfert (eds.): *Scenario Studies for the Rural Environment.* 1995   ISBN 0-7923-3748-4
6. R.E. Munn, J.W.M. la Rivière and N. van Lookeren Campagne: *Policy Making in an Era of Global Environmental Change.* 1996   ISBN 0-7923-3872-3
7. F. Oosterhuis, F. Rubik and G. Scholl: *Product Policy in Europe: New Environmental Perspectives.* 1996   ISBN 0-7923-4078-7
8. J. Gupta: *The Climate Change Convention and Developing Countries: From Conflict to Consensus?* 1997   ISBN 0-7923-4577-0
9. M. Rolén, H. Sjöberg and U. Svedin (eds.): *International Governance on Environmental Issues.* 1997   ISBN 0-7923-4701-3
10. M.A. Ridley: *Lowering the Cost of Emission Reduction: Joint Implementation in the Framework Convention on Climate Change.* 1998   ISBN 0-7923-4914-8
11. G.J.I. Schrama (ed.): *Drinking Water Supply and Agricultural Pollution.* Preventive Action by the Water Supply Sector in the European Union and the United States. 1998   ISBN 0-7923-5104-5
12. P. Glasbergen: *Co-operative Environmental Governance: Public-Private Agreements as a Policy Strategy.* 1998   ISBN 0-7923-5148-7; Pb 0-7923-5149-5
13. P. Vellinga, F. Berkhout and J. Gupta (eds.): *Managing a Material World.* Perspectives in Industrial Ecology. 1998   ISBN 0-7923-5153-3; Pb 0-7923-5206-8
14. F.H.J.M. Coenen, D. Huitema and L.J. O'Toole, Jr. (eds.): *Participation and the Quality of Environmental Decision Making.* 1998   ISBN 0-7923-5264-5
15. D.M. Pugh and J.V. Tarazona (eds.): *Regulation for Chemical Safety in Europe: Analysis, Comment and Criticism.* 1998   ISBN 0-7923-5269-6
16. W. Østreng (ed.): *National Security and International Environmental Cooperation in the Arctic – the Case of the Northern Sea Route.* 1999   ISBN 0-7923-5528-8
17. S.V. Meijerink: *Conflict and Cooperation on the Scheldt River Basin.* A Case Study of Decision Making on International Scheldt Issues between 1967 and 1997. 1999
         ISBN 0-7923-5650-0
18. M.A. Mohamed Salih: *Environmental Politics and Liberation in Contemporary Africa.* 1999   ISBN 0-7923-5650-0
19. C.J. Jepma and W. van der Gaast (eds.): *On the Compatibility of Flexible Instruments.* 1999   ISBN 0-7923-5728-0
20. M. Andersson: *Change and Continuity in Poland's Environmental Policy.* 1999
         ISBN 0-7923-6051-6

ENVIRONMENT & POLICY

ENVIRONMENT & POLICY

38. G.J.I. Schrama and S. Sedlacek (eds.): *Environmental and Technology Policy in Europe. Technological Innovation and policy integration.* 2003

ISBN 1-4020-1583-6

For further information about the series and how to order, please visit our Website
http://www.wkap.nl/series.htm/ENPO

KLUWER ACADEMIC PUBLISHERS – DORDRECHT / BOSTON / LONDON